Lecture Notes in Physics

New Series m: Monographs

The Editorial Policy for Monographs

The series Lecture Notes in Physics reports new developments in physical research and teaching - quickly, informally, and at a high level. The type of material considered for publication in the New Series m includes monographs presenting original research or new angles in a classical field. The timeliness of a manuscript is more important than its form, which may be preliminary or tentative. Manuscripts should be reasonably self-contained. They will often present not only results of the author(s) but also related work by other people and will provide sufficient motivation, examples, and applications.
The manuscripts or a detailed description thereof should be submitted either to one of the series editors or to the managing editor. The proposal is then carefully refereed. A final decision concerning publication can often only be made on the basis of the complete manuscript, but otherwise the editors will try to make a preliminary decision as definite as they can on the basis of the available information.
Manuscripts should be no less than 100 and preferably no more than 400 pages in length. Final manuscripts should preferably be in English, or possibly in French or German. They should include a table of contents and an informative introduction accessible also to readers not particularly familiar with the topic treated. Authors are free to use the material in other publications. However, if extensive use is made elsewhere, the publisher should be informed.
Authors receive jointly 50 complimentary copies of their book. They are entitled to purchase further copies of their book at a reduced rate. As a rule no reprints of individual contributions can be supplied. No royalty is paid on Lecture Notes in Physics volumes. Commitment to publish is made by letter of interest rather than by signing a formal contract. Springer-Verlag secures the copyright for each volume.

The Production Process

The books are hardbound, and quality paper appropriate to the needs of the author(s) is used. Publication time is about ten weeks. More than twenty years of experience guarantee authors the best possible service. To reach the goal of rapid publication at a low price the technique of photographic reproduction from a camera-ready manuscript was chosen. This process shifts the main responsibility for the technical quality considerably from the publisher to the author. We therefore urge all authors to observe very carefully our guidelines for the preparation of camera-ready manuscripts, which we will supply on request. This applies especially to the quality of figures and halftones submitted for publication. Figures should be submitted as originals or glossy prints, as very often Xerox copies are not suitable for reproduction. In addition, it might be useful to look at some of the volumes already published or, especially if some atypical text is planned, to write to the Physics Editorial Department of Springer-Verlag direct. This avoids mistakes and time-consuming correspondence during the production period.
As a special service, we offer free of charge LaTeX and TeX macro packages to format the text according to Springer-Verlag's quality requirements. We strongly recommend authors to make use of this offer, as the result will be a book of considerably improved technical quality. The typescript will be reduced in size (75% of the original). Therefore, for example, any writing within figures should not be smaller than 2.5 mm.
Manuscripts not meeting the technical standard of the series will have to be returned for improvement.
For further information please contact Springer-Verlag, Physics Editorial Department II, Tiergartenstrasse 17, W-6900 Heidelberg, FRG.

Alberto Lerda

Anyons

Quantum Mechanics of Particles with Fractional Statistics

Springer-Verlag Berlin Heidelberg GmbH

Author

Alberto Lerda
Dipartimento di Fisica Teorica, Universitá di Torino
Via P. Giuria, 1, I-10125 Torino, Italy

Institute for Theoretical Physics
State University of New York at Stony Brook
Stony Brook, New York 11794-3840, USA

ISBN 978-3-662-13898-4 ISBN 978-3-540-47466-1 (eBook)
DOI 10.1007/978-3-540-47466-1

Originally published by Springer-Verlag Berlin Heidelberg New York in 1992
Softcover reprint of the hardcover 1st edition 1992

Typesetting: Camera-ready by author/editor using the $\mathrm{T_EX}$ macro package from Springer-Verlag Berlin Heidelberg GmbH.

58/3140-543210 - Printed on acid-free paper

Dedication

A Marialuisa

Preface

This book is the result and the consequence of a series of seminars and lectures I gave on the subject of anyons in many places. Everything started in September 1990 at the meeting "Condensed Matter and High Energy Physics (Fractional Statistics)" in Chia (Cagliari, Italy) organized by Professors S. Fubini and L. Alvarez-Gaumè, who invited me to speak. Then, in the spring of 1991, I had the pleasure of giving a series of lectures on anyons at the Centro Brasileiro de Pesquisas Fisicas in Rio de Janeiro (Brazil) at the invitation of Professors J.A. Mignaco and M.A. do Rego Monteiro. Finally I had the opportunity of presenting part of the material of this book in a series of seminars at the Center for Theoretical Physics of M.I.T. (Cambridge, USA) and at the Institute for Theoretical Physics of S.U.N.Y. at Stony Brook (New York, USA). All of these, together with the many talks I gave on the subject, have been extremely valuable and challenging experiences for me. In this book I have tried to collect together and include as much as I have been able of all the suggestions, comments and observations that were made during my lectures, and also I have tried to answer the questions that were posed to me on those occasions.

Obviously I do not presume to have done a perfect job; I am the first to realize that my book does not cover the whole subject and that many interesting topics have been left out from my exposition. This is due partly to lack of time, and partly also to my lack of familiarity with those topics. However, as a partial justification, I would like to mention that many excellent reviews on these subjects are available, and hence a further duplication would have been superfluous. Among these reviews, to which I refer the interested reader, I would like to mention:

- D.P. Arovas, in *Geometric Phases in Physics* edited by A. Shapere and F. Wilczek (World Scientific, Singapore 1989);
- A.P. Balachandran, E. Ercolessi, G. Morandi and A.M. Srivastava, *Hubbard Model and Anyon Superconductivity* (World Scientific, Singapore 1990);
- G.S. Canright and S.M. Girvin, Science **247** (1990) 1197;
- S. Forte, Rev. Mod. Phys. **64** (1992) 193;
- P. de Sousa Gerbert, Int. Jour. Mod. Phys. **A6** (1991) 173;
- G.A. Goldin and D. H. Sharp, Int. Jour. Mod. Phys. **B5** (1991) 2625;
- R. Iengo and K. Lechner, Phys. Rep. **213** (1992) 179;
- J.D. Lykken, J. Sonnenschein and N. Weiss, Int. Jour. Mod. Phys. **A6** (1991) 5155;

- M. Stone, Int. Jour. Mod. Phys. **B4** (1990) 1465;
- X.G. Wen and A. Zee, Santa Barbara preprint, NSF-ITP-89-155 (November 1989)
- F. Wilczek, in *Fractional Statistics and Anyon Superconductivity* edited by F. Wilczek (World Scientific, Singapore 1989);
- F. Wilczek, Int. Jour. Mod. Phys. **B5** (1991) 1273.

Acknowledgements

It is a great pleasure for me to thank all my colleagues who through extensive discussion greatly enhanced my knowledge and understanding of the subject of this book. In particular I would like to mention L. Alvarez-Gaumè, M. Caselle, C. Chou, S. Forte, M. Frau, D.Z. Freedman, S. Fubini, C.A.P. Galvao, V. Korepin, R. Jackiw, J. Jain, J.G. McCarthy, J.A. Mignaco, I. Pesando, M.A. do Rego Monteiro, M. Roček, J.J.M. Verbaarschot, W.I. Weisberger and G. Zemba.

I would particularly like to thank Robert L. Jaffe for his suggestion to write these lecture notes and his constant support during this long enterprise.

Special thanks are also due to Alfred S. Goldhaber for his careful reading of the manuscript and his enlightening remarks that helped to clarify some delicate points.

I am deeply indebted to my collaborators Gerald V. Dunne and Carlo A. Trugenberger with whom I had the pleasure of doing research on anyons.

And last but not least, these lecture notes would not have been possible without the constant help and guidance of my friend, mentor, and long-time collaborator Stefano Sciuto, who introduced me to the subject of anyons and also undertook the heavy enterprise of reading and correcting my manuscript. I am deeply grateful to him for all these things and many others.

Needless to say that all the errors and misprints that are left in the book are exclusively my responsi bility.

I would like to acknowledge partial financial support from the National Science Foundation under grant NSF PHY 90-08936 with the Institute for Theoretical Physics of S.U.N.Y. at Stony Brook (New York, USA), and from the Istituto Nazionale di Fisica Nucleare, Sezione INFN di Torino (Italy). I would like to thank also the Laboratoire de Physique Théorique of ENS-LAPP (Lyon, France) for the hospitality extended to me during the very last stage of this work.

Contents

1. Introduction

Since the early days of quantum mechanics, it has been well-known that the physical behavior of collections of identical particles is influenced not only by conventional interactions but also by the statistics of the particles. These are classified into bosons and fermions depending on whether they obey the Bose-Einstein or the Fermi-Dirac statistics respectively. The particle statistics determines the structure of the many-body wavefunctions, that turn out to be symmetric under permutations of identical bosons, but antisymmetric under permutations of identical fermions.

The distinction between symmetric and antisymmetric wavefunctions, or equivalently between bosons and fermions, is as old as quantum mechanics, and has many important and far reaching consequences. For example, the Pauli exclusion principle according to which two identical fermions cannot occupy the same quantum state, naturally follows from the antisymmetry of the fermionic wavefunctions. By contrast, the probability for bosons to be in some state is enhanced when that state is already occupied, so that the ground state of a system of identical bosons is realized when all particles condense into the lowest energy state.

For many years till quite recently, bosons and fermions have been thought to be the only logical possibilities. This is certainly correct for particles moving in at least three space dimensions, but for systems in two dimensions the situation is more interesting. Here the quantum statistics is not limited to the Bose-Einstein and the Fermi-Dirac cases, but rather a continuous interpolation between these two extremes is possible. Therefore, in two space dimensions we do not have only bosons and fermions, but also particles with any statistics in between. These particles are called *anyons* and are the subject of this book.

According to the spin-statistics connection, bosons are particles with integer spin while fermions are particles with half-integer spin. Extending this idea, one may conjecture that anyons are characterized by a fractional spin, or more generally by fractional quantum numbers. As we will see hereinafter, this is indeed the case. However, the appearance of unusual quantum numbers is not an entirely new feature of an anyonic theory. In fact, long ago T.H.R. Skyrme showed that in the non-linear sigma model of pions – particles obeying Bose statistics – there are soliton configurations which carry the quantum numbers of the nucleons – particles obeying Fermi statistics (Skyrme 1961). Actually these solitons are true fermions (Finkelstein and Rubistein 1968), despite the fact that they arise from bosonic fields. Similarly, magnetic monopoles in three space dimensions can be fermions even when the elementary fields are bosons (Hasenfratz and 't Hooft 1976), and can carry fractional (Jackiw and Rebbi 1976) or even irrational electric charges (Witten 1979).

These examples show that one can trasmute the statistics and form composite objects which obey different rules from their constituents. In three dimensions, however, there is not much choice since only bosons and fermions are possible. On the contrary in two dimensions, as we have mentioned, there is a whole variety of different possibilities, and thus one can form objects obeying exhotic fractional statistics. It was only quite recently that actually these new possibilities for spin and statistics were understood. In the late 70's, J.M. Leinaas and J. Myrheim clearly stated the principles involved in the concept of fractional statistics (Leinaas and Myrheim 1977), and identified its origin in the peculiar topological properties of the configuration space of collections of identical particles. In fact, this space is multiply connected in two dimensions but only doubly connected in three or more dimensions. Furthermore, in two dimensions the fundamental group of this configuration space is the braid group (Artin 1926, 1947) which replaces the more familiar permutation group when discussing statistics. A few years later, G.A. Goldin, R. Menikoff and D.H. Sharp reached similar conclusions using a completely different method based on the rigorous study of the unitary representations of current algebras and diffeomorphism groups (Goldin *et al.* 1980, 1981) [1].

Most of the great interest that anyons have attracted in the past few years derives from the (unexpected) applications of these ideas to certain two-dimesional condensed matter systems, most notably those exhibiting the fractional quantum Hall effect (see for instance (Prange and Girvin 1990)). In this case a series of new states of matter emerge as incompressible quantum liquids (Laughlin 1983) around which the low-energy excitations are localized quasi-particles with unusual fractional quantum numbers, *i.e. anyons.* Furthermore, it is also very likely that anyonic excitations with fractional statistics exist in films of liquid ^{3}He in the A-phase (Volovik and Yakovenko 1989). The application of anyons to the theory of high temperature superconductivity has also been considered quite extensively (for reviews see (Wilczek 1990; Lykken *et al.* 1991)), but their actual relevance in this context is quite controversial and doubtful.

In this book I will consider the subject of anyons from many different points of view. In Chapter 2 I will give a general introduction to fractional statistics in two dimensions and discuss its relation to the braid group. In Chapter 3 I will show that that anyonic statistics can be implemented on ordinary bosons or fermions with the addition of special topological interactions, and in particular I will consider the connection between anyons and Chern-Simons field theories. In Chapter 4 I will show how to describe anyons with multi-valued wavefunctions carrying a representation of the braid group. Chapter 5 will be devoted to a discussion of the second-quantized theory of anyons and to the spin-statistic connection. In Chapter 6 I will consider the case of anyons in a magnetic field, which is relevant for the applications to the fractional quantum Hall effect, and derive the energy spectrum and the wavefunctions. In Chapter 7 I will address the important problem of the statistical mechanics of systems of many anyons and discuss their partition functions, their virial coefficients and their magnetic moments. Finally, in Chapters

[1] In this book we will not discuss this approach and we refer the reader to the original literature as well as to the review article (Goldin and Sharp 1991).

8 and 9 I will consider the relevance of anyons to the theory of the fractional quantum Hall effect and comment on the surprising relation between anyons and conformal field theory.

My exposition certainly does not cover the whole subject of anyons; in particular I will not consider at all the mechanisms of anyon superconductivity, the non-abelian extensions of fractional statistics and the gravitational anyons. Moreover, the topics I treated in Chapters 7 and 9 are still the subject of active current investigations, and thus my presentation is not conclusive. The list of references I have included, contains the bibliography which is directly relevant to my exposition; though quite extended, it is by no means complete and exhaustive, and I sincerely apologize for all omissions, which however are completely unintentional.

2. Introduction to Fractional Statistics in Two Dimensions

...
So many gods, so many creeds,
So many paths that wind and wind,
...

Ella Wheeler Wilcox

In two space dimensions there are possibilities for spin and statistics which do not occur in higher dimensions: the spin is not quantized in integer or half-integer units, and particles are not necessarily bosons or fermions. Indeed they may obey *any* statistics and thus are called *anyons* (Wilczek 1982a,b, 1990). These new possibilities for spin and statistics arise from the peculiar topological properties of the configuration space of collections of identical two-dimensional particles (Leinaas and Myrheim 1977; Goldin *et al.* 1980, 1981). However anyons are not simply topological fantasies or objects of purely mathematical interest; on the contrary they might play an important role in certain physical phenomena of the real world. Of course, since we are living in at least three space dimensions where particles can be only bosons or fermions, anyons are not real particles. However there exist certain condensed-matter systems (for example thin layers at the interface between different semiconductors) that can be regarded effectively as two-dimensional. Their localized excitations (if they exist) are *quasi-particles* subject to the rules of a two-dimensional world. It is these quasi-particles that may be anyons and may be observed in certain cases. For example the collective excitations above the ground state of systems exhibiting the fractional quantum Hall effect (for a review see (Prange and Girvin 1990)) have been identified as localized quasi-particles of fractional charge (Laughlin 1983), fractional spin and fractional statistics (Arovas *et al.* 1984; Halperin 1984), and thus they can be regarded as anyons. Furthermore, anyons are conjectured to play a role also in the theory of high temperature superconductivity (Chen *et al.* 1989), even though in this case no conclusive word can be said at the moment (Lyons *et al.* 1990; Kiefl *et al.* 1990; Spielman *et al.* 1990).

The fact that in two dimensions the spin may be arbitrary (that is not only integer or half-integer in units of $\hbar$) is not too surprising. In fact, the rotation group in $d = 2$ is $SO(2)$, which is an abelian group. Therefore, there are no commutation relations to quantize, and no restrictions on the possible eigenvalues of the spin come from the algebra. If we believe in some connection between spin and statistics, the arbitrariness of the spin leads us to conceive the idea that also the statistics may be arbitrary in two dimensions [2]. As we will see, this is indeed the case.

The notion of statistics is usually related to the sign that a many-body wavefunction acquires when any two particles are interchanged. If the wavefunction gets a plus sign, we say that it describes a bosonic system; if it gets a minus sign we say that it describes a fermionic system. Actually, we can give a more general definition of statistics: let $\psi(1,2)$ be the wavefunction describing two identical hard-core particles with definite angular momentum, and let us assume that when we move particle 2 around particle 1 by an azimuthal angle $\Delta\varphi$ (see Fig. 2.1), the wavefunction changes according to

$$\psi(1,2) \longrightarrow \psi'(1,2) = \mathrm{e}^{\mathrm{i}\,\nu\,\Delta\varphi}\,\psi(1,2) \quad . \tag{2.1}$$

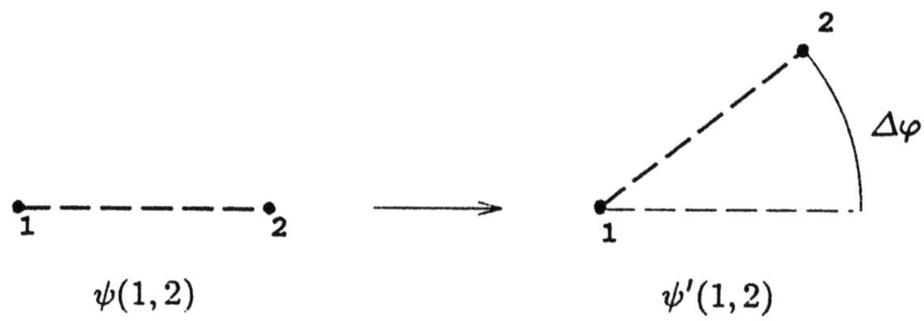

Fig. 2.1. When particle 2 is moved around particle 1 by an angle $\Delta\varphi$, the wavefunction ψ changes into ψ'.

The phase acquired by the wavefunction depends on a parameter ν which is usually called *statistics*. The meaning of ν and of (2.1) becomes more clear if we consider the exchange of the two particles. This can be realized in two ways:

i) Moving particle 2 around particle 1 by an angle $\Delta\varphi = \pi$, and then performing a rigid translation of the center of mass to reach the initial spatial configuration (see Fig. 2.2);

ii) Moving particle 2 around particle 1 by an angle $\Delta\varphi = -\pi$ and then performing the rigid translation of the center of mass (see Fig. 2.3).

In the first case the wave function acquires a phase $\exp(\mathrm{i}\pi\nu)$ according to (2.1), whilst in the second case, it gets a phase $\exp(-\mathrm{i}\pi\nu)$. This simple example shows that there is a dramatic difference between two and three or more dimensions as far as statistics is concerned. In fact in $d \geq 3$ there is no intrinsic difference between case *i*) and case *ii*) since, as is clear from the figures, we can always deform the transformation in *i*) into the one in *ii*) in a continuous way: for example we can lift the path in *i*) into the third dimension, then fold it back onto the plane, and finally superpose it to the path in *ii*). That we cannot distinguish between *i*) and *ii*) is a reflection of the fact that the azimuthal angle $\Delta\varphi$ is not a well-defined

[2] See Chapters 3 and 5 for a discussion on the spin-statistics connection for anyons.

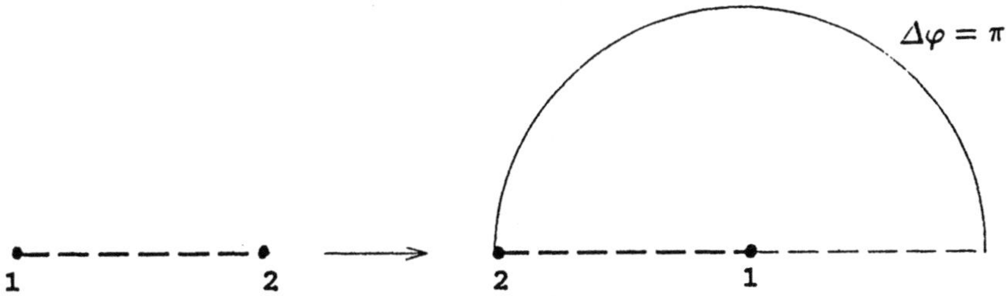

Fig. 2.2. The exchange of two particles is realized by moving particle 2 counterclockwise around particle 1 by an angle $\Delta\varphi = \pi$.

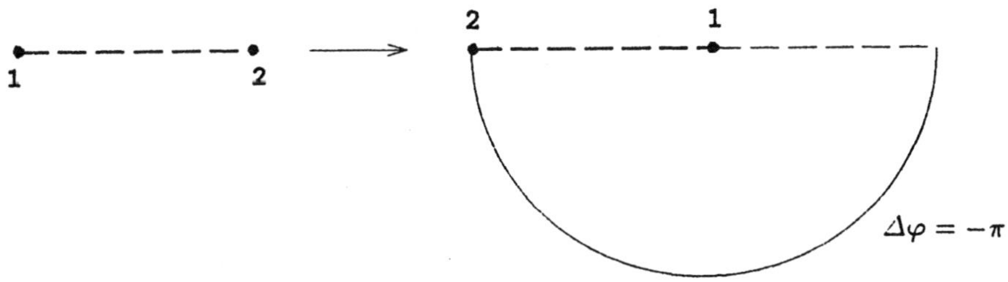

Fig. 2.3. The exchange of two particles is realized by moving particle 2 clockwise around particle 1 by an angle $\Delta\varphi = -\pi$.

quantity in $d \geq 3$. The consequences of this observation are very important: if *i*) and *ii*) are the same physical operation, the wave function should behave in the same way, which means that in $d \geq 3$ one must have

$$e^{i\pi\nu} = e^{-i\pi\nu} \quad . \tag{2.2}$$

Clearly (2.2) can only be true if $\nu = 0, 1$ (modulo 2). This simple and intuitive example shows that in $d \geq 3$ the statistics cannot be arbitrary. Under the exchange of two particles, the wavefunction picks up either a plus sign if $\nu = 0$ (bosonic statistics) or a minus sign if $\nu = 1$ (fermionic statistics). There are no other possibilities.

The situation changes drastically in two dimensions where it is not any more possible to deform continuously the path in *i*) into the one in *ii*) since by assumption the particles cannot go through each other. Hence in $d = 2$, *i*) and *ii*) are two topologically and physically distinct operations – a reflection of the fact that the azimuthal angle $\Delta\varphi$ is a well-defined quantity in two dimensions. The equality (2.2) does not necessarily hold any more and the statistical parameter ν can be arbitrary, at least in principle [3]. From this example we learn another important fact: in $d = 2$ it is not enough to specify the initial and final configurations to completely characterize a system; it is also necessary to specify how the different trajectories wind or *braid* around each other. In other words the time-evolution of the particles is important and cannot be neglected in $d = 2$. As we will see later, this fact implies that in order to classify and characterize anyons, the representations of the permutation group must be replaced by those of the more complicated *braid group*.

One of the essential features of anyons is the violation of the discrete symmetries of parity P and time reversal T. This is particularly evident in our two-particle example. In fact, under a parity or time reversal transformation, the phase acquired by the wave function $\psi(1,2)$ changes according to

$$e^{\pm i\nu\pi} \longrightarrow e^{\mp i\nu\pi} \tag{2.3}$$

for cases *i*) and *ii*) respectively, and for $\nu \neq 0, 1$ a violation of P and T occurs. This breaking of the discrete symmetries is the hallmark of anyonic statistics and is the signal that experiments look for to demonstrate the presence (or absence) of anyons in two-dimensional phenomena (Halperin *et al.* 1989).

Now we try to be more rigorous and study the problem of anyonic statistics from a more formal point of view, following essentially the presentation of Y.-S. Wu (Wu 1984a,b). Let M_N^d be the configuration space of a collection of N identical hard-core particles in d dimensions and let q and q' be two arbitrary points in M_N^d. According to the standard path-integral formulation of quantum-mechanics (see for instance (Feyman and Hibbs 1965; Laidlaw and Morette-de Witt 1971)), the amplitude for the system to evolve from the configuration q at time t to the configuration q' at time t' is given by the kernel

$$K(q',t';q,t) = \langle q',t'|q,t\rangle = \int_{q(t)=q;q(t')=q'} \mathcal{D}q \;\; e^{\frac{i}{\hbar}\int_t^{t'} d\tau\; \mathcal{L}[q(\tau),\dot{q}(\tau)]} \tag{2.4}$$

where $\mathcal{L}(q,\dot{q})$ is the lagrangian density for the N-particle system and the symbol $\int_{q(t)=q;q(t')=q'} \mathcal{D}q$ denotes the sum over all paths connecting q at time t to q' at time t'. The kernel $K(q',t';q,t)$ evolves the single-valued wavefunction $\psi(q,t)$ according to

$$\begin{aligned} \psi(q',t') &= \int_{M_N^d} dq \;\langle q',t'|q,t\rangle\langle q,t|\psi\rangle \\ &= \int_{M_N^d} dq \; K(q',t';q,t)\;\psi(q,t) \quad . \end{aligned} \tag{2.5}$$

Without any loss of generality, we choose $q = q'$ and hence describe loops in M_N^d. Two loops are considered equivalent (or homotopic) if one can be obtained from the other by a continuous deformation. All homotopic loops are grouped

[3] We will see that actually there are restrictions on ν coming from the topology of the two dimensional space. For example for particles moving on a torus, ν can be only a rational number.

into one class and the set of all such classes is called the fundamental group [4] and is denoted by π_1. Thus an element of $\pi_1\left(M_N^d\right)$ is simply the set of all loops in M_N^d which can be continuously deformed into each other. On the other hand, loops belonging to two different elements of $\pi_1\left(M_N^d\right)$ cannot be connected by a continuous transformation. With this in mind we can organize the sum over all loops in (2.4) into a sum over homotopic classes $\alpha \in \pi_1\left(M_N^d\right)$ and into a path-integral in each class. Therefore (2.4) may be rewritten as

$$\begin{aligned} K(q,t';q,t) &= \sum_{\alpha\in\pi_1\left(M_N^d\right)} K_\alpha(q,t';q,t) \\ &= \sum_{\alpha\in\pi_1\left(M_N^d\right)} \int_{q_\alpha(t)=q;q_\alpha(t')=q} \mathcal{D}q_\alpha \; e^{\frac{i}{\hbar}\int_t^{t'} d\tau\; \mathcal{L}[q_\alpha(\tau),\dot{q}_\alpha(\tau)]} \quad . \end{aligned} \tag{2.6}$$

This formula can be interpreted as a decomposition of the amplitude $K(q,t';q,t)$ into a sum of subamplitudes $K_\alpha(q,t';q,t)$ to which only homotopic loops contribute. With such a decomposition, it is clear that in principle we can assign different weights to the different subamplitudes $K_\alpha(q,t';q,t)$ provided that we preserve the conventional rules for the composition of probabilities. Thus, instead of (2.6) we can write

$$K(q,t;q,t') = \sum_{\alpha\in\pi_1\left(M_N^d\right)} \chi(\alpha) \int_{q_\alpha(t)=q;q_\alpha(t')=q} \mathcal{D}q_\alpha \; e^{\frac{i}{\hbar}\int_t^{t'} d\tau\; \mathcal{L}[q_\alpha(\tau),\dot{q}_\alpha(\tau)]} \tag{2.7}$$

where $\chi(\alpha)$ are some complex numbers. For (2.7) to make sense as a probability amplitude, the weights $\chi(\alpha)$ cannot be arbitrary. In fact, since we want to mantain the usual rule for combining probabilities

$$\begin{aligned} K(q'',t'';q,t) &= \langle q'',t''|q,t\rangle \\ &= \int_{M_N^d} dq' \; \langle q'',t''|q',t'\rangle\langle q',t'|q,t\rangle \\ &= \int_{M_N^d} dq' \; K(q'',t'';q',t') \; K(q',t';q,t) \quad , \end{aligned} \tag{2.8}$$

the weights $\chi(\alpha)$ must satisfy

$$\chi(\alpha_1)\chi(\alpha_2) = \chi(\alpha_1\cdot\alpha_2) \tag{2.9}$$

for any α_1 and α_2. Equation (2.9) can also be read as the statement that $\chi(\alpha)$ must be a one-dimensional representation of the fundamental group $\pi_1\left(M_N^d\right)$.

To see which representations are possible, we have to specify better what are M_N^d and its fundamental group. To this end, let us consider a system of N identical

---

[4] In the set π_1 one can define a product $\cdot$ in a very simple and natural way: if α_1 and α_2 are two classes with representatives q_1 and q_2 respectively, then $\alpha_1\cdot\alpha_2$ is the class whose representative is the loop q_1q_2 (that is the loop q_1 followed by the loop q_2). It can be shown that this product furnishes π_1 with a group structure.

hard-core particles moving in the euclidean d-dimensional space, $\mathbb{R}^d$. A configuration of such a system is clearly specified by the N coordinates of the particles, that is by an element of $(\mathbb{R}^d)^N$. However because of the hard-core assumption [5], any two particles cannot occupy the same position, which means that from $(\mathbb{R}^d)^N$ we have to remove the generalized diagonal

$$\Delta = \left\{ (\boldsymbol{r}_1, \ldots, \boldsymbol{r}_N) \in (\mathbb{R}^d)^N : \boldsymbol{r}_I = \boldsymbol{r}_J \text{ for some } I \neq J \right\} \quad . \tag{2.10}$$

Furthermore our particles are identical and indistinguishable, and thus we should identify configurations which differ only in the ordering of the particles. In other words we should divide by the permutation group S_N. Therefore we conclude that the configuration space for our system is

$$M_N^d = \frac{(\mathbb{R}^d)^N - \Delta}{S_N} \quad . \tag{2.11}$$

To find the fundamental group of such a space is a standard problem in algebraic topology, which was addressed and solved in the early 60's (Fadell and Neuwirth 1962; Fox and Neuwirth 1962; Fadell and Van Buskirk 1962). Here we simply quote the results and refer the reader to the specialized literature for their derivation. It turns out that the fundamental group of M_N^d is given by

$$\pi_1\left(M_N^d\right) = \begin{cases} S_N & \text{if } d \geq 3 \\ B_N & \text{if } d = 2 \end{cases} \tag{2.12}$$

where B_N is Artin's braid group of N objects which contains the permutation group S_N as a finite subgroup (Artin 1926, 1947).

Even from this formal point of view we see again that there is a crucial difference between two and three or more dimensions, exactly as we discovered in the simple example at the beginning of this chapter. To have a more explicit understanding of (2.12), let us reconsider our two-particle example in the light of what we have just observed, following the discussion of (Leinaas and Myrheim 1977). Let us start with the case of two dimensions. Instead of assigning the position vectors $\boldsymbol{r}_1$ and $\boldsymbol{r}_2$ for the two particles, it is more convenient to introduce the center-of-mass coordinate

$$\boldsymbol{R} = \frac{1}{2}(\boldsymbol{r}_1 + \boldsymbol{r}_2) \quad \in \mathbb{R}^2 \quad , \tag{2.13}$$

and the relative coordinate

$$\boldsymbol{r} = \boldsymbol{r}_1 - \boldsymbol{r}_2 \quad \in \mathbb{R}^2 - \{0\} \quad . \tag{2.14}$$

(We have removed the origin because of the hard-core requirement $\boldsymbol{r}_1 \neq \boldsymbol{r}_2$.) Since $\boldsymbol{R}$ is invariant under the permutations of S_2, we can write

$$M_2^2 = \mathbb{R}^2 \times r_2^2 \tag{2.15}$$

[5] It will be clear later on why this assumption is important to discuss non trivial statistics.

where r_2^2 is "some" space describing the two degrees of freedom of the relative motion. We now argue that r_2^2 has the topology of a cone. Since two configurations which differ only in the ordering of the particle indices are indistinguishable, $\boldsymbol{r}$ and $-\boldsymbol{r}$ must be identified. The space r_2^2 is then the upper-half plane without the origin and with the positive x-axis identified with the negative one. The results of such identification is a cone without tip (see Fig. 2.4).

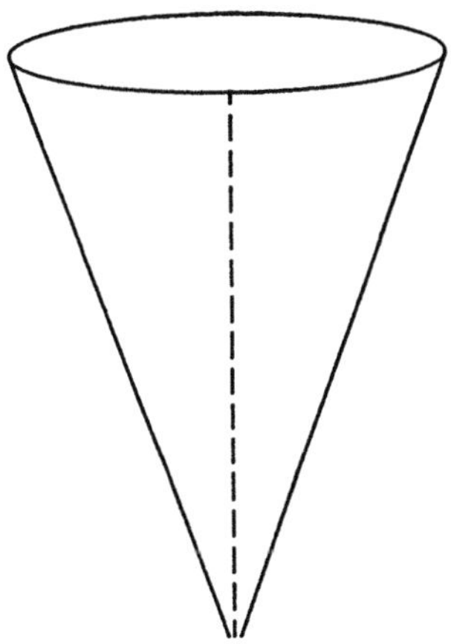

Fig. 2.4. The manifold r_2^2 describing the relative coordinate of two identical particles in two dimensions has the topology of a cone without the tip.

According to the decomposition (2.15), any loop in M_2^2 can be classified by the number of times it winds around the cone r_2^2. Two loops q and q' with different winding numbers are homotopically inequivalent: It is not possible to deform one into the other since the vertex of the cone has been removed. Thus the spaces r_2^2 and $\mathbb{R}^2 \times r_2^2$ are infinitely connected, and

$$\pi_1\left(M_2^2\right) = \pi_1\left(\mathbb{R}^2 \times r_2^2\right) = \mathbb{Z} = B_2 \quad . \tag{2.16}$$

It is important to realize that if the vertex of the cone were included – that is if we allowed particles to occupy the same position – the configuration space would be simply connected since any loop, even when winding around the cone, could be contracted to a point by deforming and unwinding it through the tip. Thus, if we do not impose the hard-core constraint on the particles, we can describe only bosonic statistics.

Let us now turn to the case of two particles in three dimensions. After introducing the center-of-mass coordinate $\boldsymbol{R} \in \mathbb{R}^3$, we can decompose the configuration space M_2^3 as

$$M_2^3 = \mathbb{R}^3 \times r_2^3 \tag{2.17}$$

where the space r_2^3 describes the three degrees of freedom of the relative motion. These are the length and the two angles of the relative coordinate $\boldsymbol{r}$, and as before

$\boldsymbol{r}$ and $-\boldsymbol{r}$ are identified. It is easy to realize that r_2^3 is just the product of the semi-infinite line describing $|\boldsymbol{r}|$ times the projective space $\mathcal{P}_2$ describing the orientation of $\pm\boldsymbol{r}/|\boldsymbol{r}|$. In turn $\mathcal{P}_2$ can be described as the northern hemisphere with opposite points on the equator being identified (see Fig. 2.5).

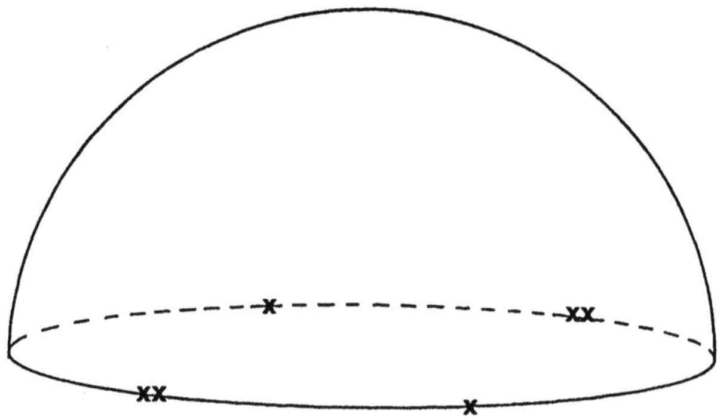

Fig. 2.5. The projective space $\mathcal{P}_2$ can be represented as the northern hemisphere with opposite points on the equator being identified.

The space $\mathcal{P}_2$ is doubly connected and admits only two classes of loops: Those which can be shrunk to a point by a continuous transformation and those which cannot. In Fig. 2.6 we exhibit a typical contractible loop q_1 and a typical non-contractible loop q_2.

There are no other *classes because the square of a non-contractible loop is* contractible as Fig. 2.7 shows in a specific example.

Therefore from the decomposition (2.17) and the topology of r_2^3, we deduce that

$$\pi_1\left(M_2^3\right) = \pi_1\left(\mathbb{R}^3 \times r_2^3\right) = \mathbb{Z}_2 = S_2 \quad . \tag{2.18}$$

Thus only bosons and fermions can exist, the former corresponding to contractible loops and the latter to non-contractible loops.

We have seen that at the heart of anyonic statistics there is the braid group B_N in place of the permutation group S_N which is responsible for ordinary statistics. In fact, there are just two one-dimensional representations of S_N (the identical one and the alternating one, corresponding respectively to bosonic and fermionic statistics), whereas the braid group B_N admits a whole variety of one-dimensional representations whose labelling parameter will be identified with the statistics ν.

A few remarks are in order at this point. So far we have been focusing on one-dimensional representations of $\pi_1\left(M_N^d\right)$ because we are considering the simplest possible case of scalar quantum mechanics, when the wavefunctions are one-component objects. Of course we could consider more complicated situations where wavefunctions are multiplets; in such cases of course appropriate higher-dimensional representations of $\pi_1(M_N)$ would be necessary. In the past few years

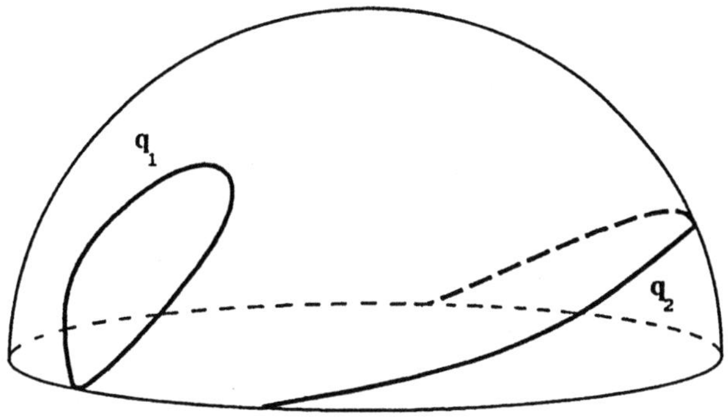

Fig. 2.6. Examples of contractible (q_1) and non-contractible (q_2) loops. Notice that even though q_2 is depicted as an open curve, it is really a closed loop because the two end points on the equator are identified in $\mathcal{P}_2$.

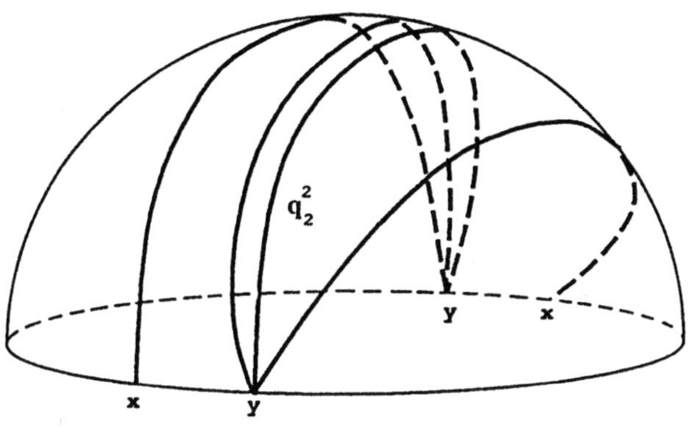

Fig. 2.7. The square of a non-contractible loop, q_2^2, is contractible. In fact, the contractible loop passing through x can be considered as a deformation of q_2^2.

higher-dimensional representations of the braid group B_N have been seen to play important roles in a variety of different contexts. Among these we can mention the study of the monodromy properties of the N-point conformal blocks in conformal field theories (Moore and Seiberg 1989); the theory of links, knots and Jones polynomials (Witten 1989); the exactly solvable models and the Yang-Baxter equation (see for instance (Yang and Ge 1989); the theory of non-abelian anyons (Moore and Read 1991). On the other hand, higher-dimensional representations of the permutation group S_N were considered long ago in the context of parastatistics. In this respect it should be clear that anyonic statistics (which is typical of two dimensions) and parastatistics (which can be defined in any dimension) are two completely different concepts, because they are based on two different structures, the braid group and the permutation group respectively. Furthermore it can be rigorously proven that parastatistics cannot be realized in a non trivial way within a *local* theory for $d > 2$ (for an extensive discussion of this result see (Fröhlich and Marchetti 1988, 1989; Fröhlich *et al.* 1989)). More simply this means that particles obeying parastatistics and having local interactions can be described as ordinary bosons or fermions with new degrees of freedom. Indeed the famous puzzle in the quantum numbers of the baryonic resonance Δ^{++} (for which the idea of parastatistics was introduced in physics) can be solved either with the assumption that quarks obey parastatistics (Greenberg 1964) or equivalently with the assumption that fermionic quarks carry a new quantum number, the color (Han and Nambu 1965; Nambu 1966).

To fully appreciate the new features of anyonic statistics as compared to ordinary statistics, let us now have a closer look at the braid group. The braid group of N strands, B_N, is an infinite group which is generated by $N-1$ elementary moves $\sigma_1, \ldots, \sigma_{N-1}$ satisfying

$$\sigma_I \ \sigma_{I+1} \ \sigma_I = \sigma_{I+1} \ \sigma_I \ \sigma_{I+1} \tag{2.19}$$

for $I = 1, \ldots, N-2$, and

$$\sigma_I \ \sigma_J = \sigma_J \ \sigma_I \tag{2.20}$$

for $|I - J| \geq 2$. The inverse of σ_I is denoted by ${\sigma_I}^{-1}$, the identity by $\mathbb{1}$, and the center of B_N is generated by $(\sigma_1 \ldots \sigma_{N-1})^N$. To describe the elementary moves σ_I, it is convenient to introduce a pictorial representation as follows. Given N vertical strands, the generator σ_I acts on them by simply braiding the I-th strand around the $(I+1)$-st in a definite way, as shown in Fig. 2.8.

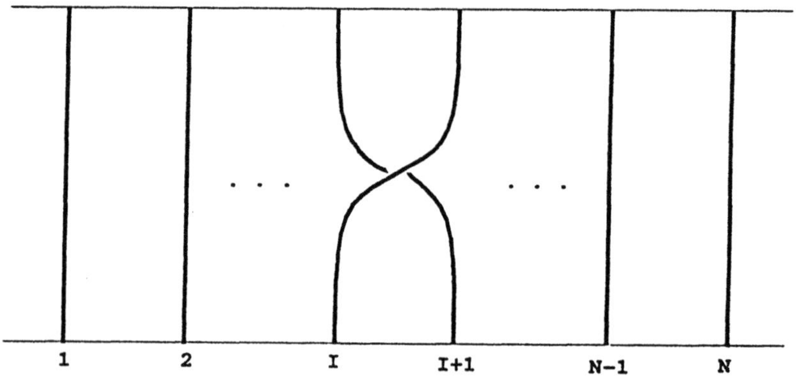

Fig. 2.8. Pictorial representation of the elementary move σ_I.

The inverse generator ${\sigma_I}^{-1}$ is represented by the move shown in Fig. 2.9.

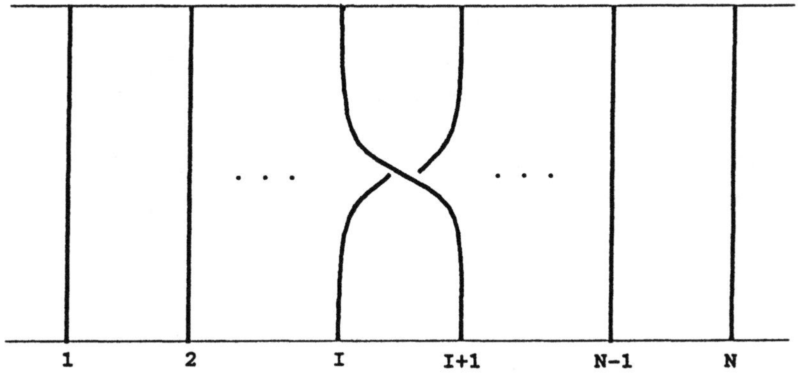

Fig. 2.9. Pictorial representation of $\sigma_I{}^{-1}$.

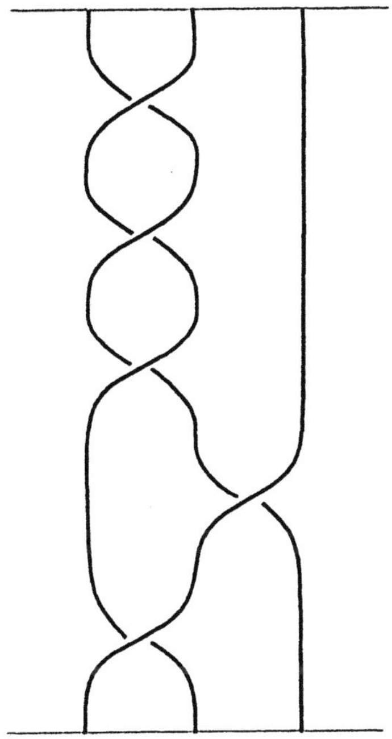

Fig. 2.10. Pictorial representation of the braid $\sigma_1^3\,\sigma_2\,\sigma_1$.

A generic braid is a word in the generators σ_I and their inverses σ_I^{-1}, which however can be rearranged using the relations (2.19) and (2.20). For example, for $N = 3$ the braid represented in Fig. 2.10 corresponds to $\sigma_1^3\,\sigma_2\,\sigma_1$ (we adopt the convention that the generators on the right act first).

We remark (and Fig. 2.10 shows it explicitly) that in general $\sigma_I{}^2 \neq 1$. If $\sigma_I{}^2 = \mathbb{1}$ for all I, then the braid group reduces simply to the permutation group S_N, which is a finite group. Diagrams like those in Figs. 2.8, 2.9, or 2.10 can also be interpreted as describing the time evolution of identical particles if we regard the strands as world-lines. To see this we can consider an easy example with three particles. Let us suppose that at time t their configuration is the one displayed in Fig. 2.11. We can describe it by listing the azimuthal angles of all possible pairs of particles measured with respect to some arbitrary reference axes. In the specific case of Fig. 2.11 we have

$$\begin{aligned} \varphi_{12}(t) &= 0 \ , \\ \varphi_{13}(t) &= \eta \ , \\ \varphi_{23}(t) &= \xi \ . \end{aligned} \tag{2.21}$$

The symbol φ_{IJ} denotes the azimuthal angle of particle J with respect to particle I, namely

$$\varphi_{IJ} = \tan^{-1}\left(\frac{x_J^2 - x_I^2}{x_J^1 - x_I^1}\right) \ , \tag{2.22}$$

(x_I^1, x_I^2) being the cartesian coordinates of the I-th particle.

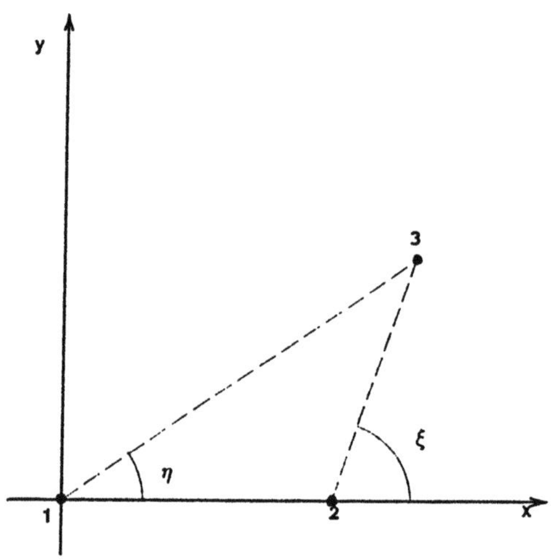

Fig. 2.11. Configuration of three particles at time t.

Let us then suppose that at time t' the particles reach the positions shown in Fig. 2.12. Now the winding angles are

$$\begin{aligned} \varphi_{12}(t') &= \xi + \pi \ , \\ \varphi_{13}(t') &= \eta + \pi \ , \\ \varphi_{23}(t') &= \pi \ . \end{aligned} \tag{2.23}$$

The configuration at time t in Fig. 2.11 and the one at time t' in Fig. 2.12 are the same even if particles 1 and 3 have been interchanged. This is because all

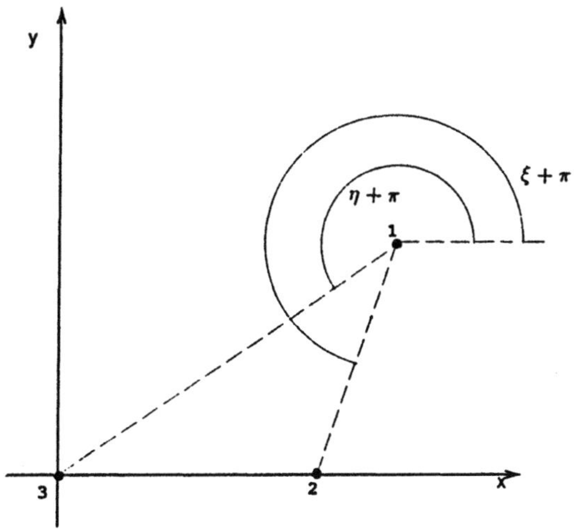

Fig. 2.12. Configuration of the three particles at time t'

particles are indistinguishable! Notice that $\varphi_{IJ}(t') - \varphi_{IJ}(t)$ is not in general an integer multiple of π; however it is always true that

$$\sum_{I<J} \varphi_{IJ}(t') - \sum_{I<J} \varphi_{IJ}(t) = n\,\pi \tag{2.24}$$

where n is an integer (in our example $n = 3$). This can be interpreted by saying that to complete a loop in configuration space an integer number of exchanges is always necessary. We can see explicitly this fact in our example if we place Fig. 2.12 on top of Fig. 2.11 and draw lines to connect the initial and final positions of each particle. In a side-view this is represented by the braiding of Fig. 2.13, which is $\sigma_1\ \sigma_2\ \sigma_1$ in terms of the elementary moves.

We see that indeed there are $n = 3$ generators, *i.e.* $n = 3$ exchanges. Notice that if we considered the braiding $\sigma_1^3\ \sigma_2\ \sigma_1$ (see Fig. 2.10) instead of $\sigma_1\ \sigma_2\ \sigma_1$, the final configuration would look the same as in Fig. 2.12, but the winding angles would be different (specifically $\varphi_{12}(t')$ would be $\xi + 3\pi$ instead of $\xi + \pi$). If we just look at the particle positions on the plane, there is no way of distinguishing between these two cases. They correspond to the same permutation of the initial positions, but if $\sigma_I^2 \neq \mathbb{1}$, they are dynamically different (compare Fig. 2.10 and Fig. 2.13). Again we rediscover that the actual braiding of the particles, and not simply their permutations, are the important things to be considered in discussing anyonic statistics.

Let us now come back to the problem of finding the one-dimensional representations χ of the braid group. These are given by

$$\chi(\sigma_K) = \mathrm{e}^{-\mathrm{i}\,\nu\,\pi} \tag{2.25}$$

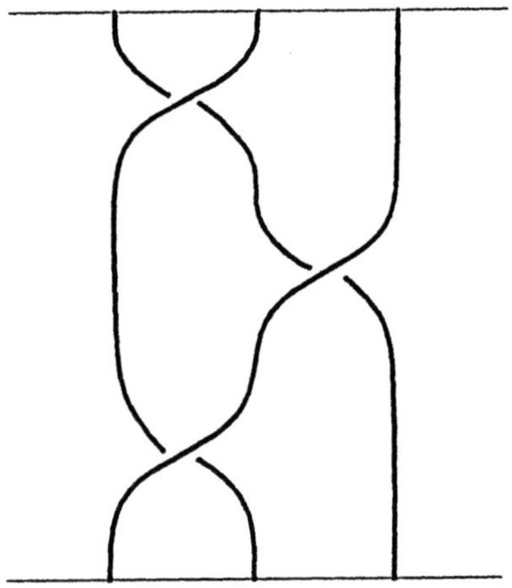

Fig. 2.13. Pictorial representation of the braid $\sigma_1\,\sigma_2\,\sigma_1$.

for any $K = 1, ..., N-1$, where ν is a real parameter defined modulo 2, which will be identified with the statistics. Since in general ${\sigma_K}^2 \neq \mathbb{1}$, ν is an arbitrary number. In the elementary move σ_K all winding angles φ_{IJ} remain constant except for $\varphi_{K,K+1}$ which changes by π. Thus we can rewrite $\chi(\sigma_K)$ as

$$\chi(\sigma_K) = \exp\left[-\mathrm{i}\,\nu\;\Delta\varphi_{K,K+1}\right] = \exp\left[-\mathrm{i}\,\nu \sum_{I<J} \Delta\varphi_{IJ}^{(K)}\right] \tag{2.26}$$

where we have introduced

$$\Delta\varphi_{IJ}^{(K)} \equiv \varphi_{IJ}^{(K)}(t') - \varphi_{IJ}^{(K)}(t) = \pi\;\delta_{I,K}\delta_{J,K+1} \tag{2.27}$$

to represent the change of the winding angles in the move σ_K. Using this notation, (2.26) can be easily generalized to an arbitrary braiding α, and hence we can write

$$\chi(\alpha) = \exp\left[-\mathrm{i}\,\nu \sum_{I<J} \int_t^{t'} d\tau\;\frac{d}{d\tau}\varphi_{IJ}^{(\alpha)}(\tau)\right] \tag{2.28}$$

where the increment $\varphi_{IJ}^{(\alpha)}(t') - \varphi_{IJ}^{(\alpha)}(t)$ has been written as an integral over an evolution parameter τ. Notice that the functions $\varphi_{IJ}^{(\alpha)}(\tau)$ are in general very complicated and can be specified only when the dynamics of the particles is fully taken into account. Therefore at the moment, we have to regard (2.28) just as a formal definition which however is extremely useful. In fact, substituting (2.28) into (2.7), we obtain

$$K(q,t';q,t) = \sum_{\alpha\in\pi_1(M_N^d)} \int_{q_\alpha(t)=q;q_\alpha(t')=q} \mathcal{D}q_\alpha \; e^{\frac{i}{\hbar}\int_t^{t'} d\tau \left\{ \mathcal{L}[q_\alpha(\tau),\dot{q}_\alpha(\tau)] - \hbar\,\nu \sum_{I<J} \frac{d\varphi_{IJ}^{(\alpha)}(\tau)}{d\tau} \right\}}$$

(2.29)

If we now define

$$\mathcal{L}' = \mathcal{L} - \hbar\,\nu \sum_{I<J} \frac{d}{d\tau}\varphi_{IJ}^{(\alpha)}(\tau) \ , \tag{2.30}$$

we see that the kernel $K(q,t;q,t')$ is decomposed with respect to $\mathcal{L}'$ into subamplitudes each of which is weighted equally as if we were describing bosons, that is

$$K(q,t';q,t) = \sum_{\alpha\in\pi_1(M_N^d)} \int_{q_\alpha(t)=q;q_\alpha(t')=q} \mathcal{D}q_\alpha \; e^{\frac{i}{\hbar}\int_t^{t'} d\tau\, \mathcal{L}'} \ . \tag{2.31}$$

In other words, instead of dealing with anyons governed by the Lagrangian $\mathcal{L}$, we can deal with bosons whose dynamics is dictated by the new Lagrangian $\mathcal{L}'$ given in (2.30). Thus, we can trade anyonic statistics for some kind of "ficticious" force and describe anyons as ordinary particles (bosons for example) with an additional statistical interaction. Notice that this statistical interaction is very peculiar and intrinsically topological in nature – in our example it is actually a total derivative. Its addition to the lagrangian $\mathcal{L}$ clearly does not change the equations of motion, which are a reflection of the local structure of the configuration space, but *does* change the statistical properties of the particles, which are instead related to the global topological structure of the configuration space. In the next chapter we will give an explicit derivation of such statistical interaction, we will clarify its topological character and see that it can be locally realized as a gauge theory with a Chern-Simons kinetic term.

So far we have considered N indistinguishable particles moving on the plane, and have seen that there exist infinitely many one-dimensional representations of the braid group B_N labeled by a continuous statistical parameter ν on which *a priori* no restrictions are made. We now briefly consider the case in which N indistinguishable particles move on a *compact* Riemann surface Σ. The corresponding braid group is

$$B_N(\Sigma) \equiv \pi_1\left(\frac{(\Sigma)^N - \Delta}{S_N}\right) \tag{2.32}$$

and has been classified for all kinds of surfaces Σ (Birman 1969; Scott 1970). As far as fractional statistics is concerned, it turns out that the topology of Σ plays a very important role in that it restricts the possible values of ν, the number of particles N and the number of components of the wavefunctions.

Let us first consider the case of the sphere ($\Sigma = S^2$). The braid group $B_N(S^2)$ is generated by σ_I with $I = 1,\ldots,N-1$ which satisfy the Artin relations (2.19) and (2.20) plus an additional constraint:

$$\sigma_1\,\sigma_2\ldots\sigma_{N-1}^2\ldots\sigma_2\,\sigma_1 = 1 \ . \tag{2.33}$$

This expresses the fact that a closed loop taking one particle around all other particles can be continuously deformed and shrunk to a point on the back side of the sphere. If we use (2.25) inside the relation (2.33), we obtain

$$e^{-i2(N-1)\nu\pi} = 1 \ , \tag{2.34}$$

which immediately restricts ν to be rational. If $\nu = p/q$ with p and q coprime integers, (2.34) is equivalent to a restriction on the number of particles, namely

$$N = qn + 1 \tag{2.35}$$

where n is a non-negative integer. Clearly for bosons ($\nu = 0$) or fermions ($\nu = 1$) there are no restrictions coming from (2.34). On Riemann surfaces of higher genus, the restrictions are even more complicated. A Riemann surface of genus g, Σ_g, can be described as a sphere with g handles (see for instance (Farkas and Kra 1980)) each of which is characterized by a pair of non-contractible curves (homology cycles). Hence it is not difficult to realize that in this case besides the exchange generators σ_I, the braid group $B_N(\Sigma_g)$ has also "topological" generators which take the particles along the $2g$ homology cycles around the handles. Following the standard notation, we call these generators $\tau_{i,I}$ and $\rho_{i,I}$ where $I = 1, ..., N-1$ and $i = 1, ..., g$. The generators σ_I, $\tau_{i,I}$ and $\rho_{i,I}$ satisfy the Artin relations (2.19) and (2.20) and

$$\tau_{i,I+1} = \sigma_I \, \tau_{i,I} \, \sigma_I^{-1} \ , \tag{2.36a}$$

$$\rho_{i,I+1} = \sigma_I \, \rho_{i,I} \, \sigma_I^{-1} \ , \tag{2.36b}$$

$$\rho_{i,I+1}^{-1} \, \tau_{i,I+1} \, \rho_{i,I+1} \, \tau_{i,I+1}^{-1} = \sigma_I^2 \ , \tag{2.36c}$$

$$\sigma_1 \, \sigma_2 \ldots \sigma_{N-1}^2 \ldots \sigma_2 \, \sigma_1 = \prod_{i=1}^{g} \rho_{i,1}^{-1} \, \tau_{i,1} \, \rho_{i,1} \, \tau_{i,1}^{-1} \ . \tag{2.36d}$$

In particular the last equation is the higher genus generalization of (2.33). We refer the reader to the specialized literature (Scott 1970; Einarsson 1990, 1991; Imbo and March-Russell 1990) for a proof and a discussion of these relations; here we limit ourselves to a brief analysis of the consequences of (2.36) on the quantum mechanics of particles of fractional statistics.

Let us consider the case of a one-dimensional unitary representation of the braid group $B_N(\Sigma_g)$, as is appropriate for scalar wavefunctions. Every generator is represented as a phase, and from (2.36c) it follows immediately that

$$1 = e^{-i2\nu\pi} \ , \tag{2.37}$$

i.e. $\nu = 0, 1$. Therefore, on a compact Riemann surface of genus $g \geq 1$ only bosons and fermions are possible if the multi-particle wavefunctions carry a one-dimensional representation of $B_N(\Sigma_g)$; this is in contrast with the case of the plane where any statistics is allowed with scalar wavefunctions.

However anyons are possible even on a compact surface provided that wavefunctions with many components are considered. In this case one has to look at higher dimensional representations of $B_N(\Sigma_g)$ which lead to the concepts of *generalized fractional statistics* and *generalized anyons* (Einarsson 1990, 1991; Imbo and March-Russell 1990; Wen *et al.* 1990a). These representations are characterized by the fact that the exchange moves σ_I are represented as a phase factor times a k-dimensional unit matrix, namely

$$\sigma_I = \mathrm{e}^{-\mathrm{i}\nu\pi}\ \mathbb{1}_k \ , \tag{2.38}$$

while the topological moves $\tau_{i,I}$ and $\rho_{i,I}$ are represented as unitary $k \times k$ matrices. The interpretation of such a representation is that the generators σ_I act on a wavefunction with k components by multiplying all components by the *same* phase factor, while the generators $\tau_{i,I}$ and $\rho_{i,I}$ mix among themselves the components of the wavefunction. Using (2.38) in (2.36a,b), we find

$$\begin{aligned} \chi(\tau_{i,I+1}) &= \chi(\tau_{i,I}) \equiv \tau_i \ , \\ \chi(\rho_{i,I+1}) &= \chi(\rho_{i,I}) \equiv \rho_i \ . \end{aligned} \tag{2.39}$$

Furthermore, (2.36c) implies

$$\rho_i^{-1}\ \tau_i\ \rho_i\ \tau_i^{-1} = \mathrm{e}^{-2\mathrm{i}\nu\pi} \ ,$$

or

$$\tau_i\ \rho_i = \mathrm{e}^{-2\mathrm{i}\nu\pi}\ \rho_i\ \tau_i \ . \tag{2.40}$$

If we now use this last relation in (2.36d), we easily get

$$\mathrm{e}^{-\mathrm{i}2(N-1)\nu\pi} = \mathrm{e}^{-2\mathrm{i}g\,\nu\pi} \tag{2.41}$$

which is the higher genus generalization of (2.34). This equation restricts ν to be a rational number. If $\nu = p/q$ with p and q coprime integers, then

$$N = q\,n + 1 - g \ . \tag{2.42}$$

For $g = 0$, we recover the sphere condition (2.35). When $\nu = p/q$ it is possible to derive explicit expressions for the $k \times k$ matrices τ_i and ρ_i. Consistency with (2.36) requires that $k = q^g$ (Einarsson 1990, 1991). Thus, on compact surfaces, only rational statistics are allowed, and the number of particles and the dimensionality of the braid group representations are not arbitrary. For an extensive discussion of anyons on compact surfaces and on the torus ($g = 1$) in particular, we refer the reader to the recent review by R. Iengo and K. Lechner (Iengo and Lechner 1992). Throughout the rest of this book, we will always consider the case of particles moving on the plane, where, as we have seen, no restrictions apply.

We conclude this introductory chapter by summarizing the main points of our discussions. The configuration space of identical two dimensional particles has a non trivial topology and is an infinitely connected space. Its fundamental group is the braid group whose representations are labeled by an arbitray parameter ν called statistics. This unusual statistics can be implemented on ordinary particles (for instance bosons) by the addition of a topological statistical interaction.

3. Fractional Statistics in the Chern-Simons Gauge

In Chapter 2 we have seen that fractional statistics can be implemented on ordinary bosons by the addition of fictitious statistical interactions (the same of course would be true on ordinary fermions). In this chapter we are going to discuss this scenario in more detail and find that the use of Chern-Simons gauge fields provides an efficient method to realize anyons. Before this however, we consider useful to discuss the dynamics of a charged point-particle interacting with an infinitely long magnetic solenoid (flux tube). This system has been called *cyon* (Goldhaber 1982) and is the prototype for anyons. When the motion along the solenoid is ignored, the dynamics is essentially planar and the cyon is subject to the laws of the two-dimensional world. In particular it may acquire fractional statistics.

3.1 The Cyon System and Its Symmetries

Let us consider a non-relativistic particle of mass m and electric charge e that moves in the magnetic field B created by an infinitely long and thin solenoid passing through the origin and directed along the z-axis. If we neglect the (free) motion along the solenoid, the relevant dynamics takes place in the (x, y)-plane and is governed by the non-relativistic Lagrangian

$$\mathcal{L} = \frac{1}{2} m v^2 + \frac{e}{c} \boldsymbol{v} \cdot \boldsymbol{A}(\boldsymbol{r}) \tag{3.1.1}$$

where $\boldsymbol{r} = (x, y) \in \mathbb{R}^2$ denotes the particle position, $\boldsymbol{v} = \dot{\boldsymbol{r}}$ its velocity and $\boldsymbol{A}$ the vector potential for the solenoid configuration. In a convenient symmetric gauge, $\boldsymbol{A}$ is given by

$$\boldsymbol{A}(\boldsymbol{r}) = \frac{\Phi}{2\pi} \left(\frac{-y}{x^2 + y^2} \hat{\boldsymbol{\imath}} + \frac{x}{x^2 + y^2} \hat{\boldsymbol{\jmath}} \right) \tag{3.1.2}$$

where $\hat{\boldsymbol{\imath}}$ and $\hat{\boldsymbol{\jmath}}$ are the unit vectors along the x- and the y-axis respectively, and Φ is the flux of the solenoid. The magnetic field B (which in two dimensions is a pseudo-scalar) associated to (3.1.2) is that of a vortex of strength Φ localized at the origin,

$$B = \nabla \wedge \boldsymbol{A} = \Phi \, \delta^{(2)}(\boldsymbol{r}) \tag{3.1.3}$$

where the δ-function arises from the singularity at $x = y = 0$ of the vector potential (3.1.2). Clearly we have

$$\int d^2r\ B = \Phi \quad , \tag{3.1.4}$$

confirming the interpretation of Φ as the flux of the solenoid. Following (Goldhaber 1982) we will call this particle-vortex system, a cyon. The canonical momentum $\boldsymbol{p}$ can be easily derived from the Lagrangian (3.1.1); it differs from the kinetic momentum $m\boldsymbol{v}$ by a term due to the vector potential $\boldsymbol{A}$, namely

$$\boldsymbol{p} = \frac{\partial \mathcal{L}}{\partial \boldsymbol{v}} = m\boldsymbol{v} + \frac{e}{c}\boldsymbol{A} \quad . \tag{3.1.5}$$

The Hamiltonian associated to (3.1.1) is

$$H = \boldsymbol{p}\cdot\boldsymbol{v} - \mathcal{L} = \frac{1}{2m}\left(\boldsymbol{p} - \frac{e}{c}\boldsymbol{A}\right)^2 = \frac{1}{2}m\boldsymbol{v}^2 \quad . \tag{3.1.6}$$

Notice that the magnetic field and the vector potential are invisible when the Hamiltonian is written in terms of the kinetic momentum. In fact, as we see in (3.1.6), the Hamiltonian of the cyon is numerically equal to that of a free particle! The only effect of the presence of the vortex is to enforce the non-trivial relationship (3.1.5) between the canonical and the kinetic momenta. The classical equations of motion are the same as those of a free particle, but non-trivial features are present in the quantum theory.

The Lagrangian (3.1.1) is rotationally invariant: If the dynamical variables are rotated in the plane, $\mathcal{L}$ does not change. The constant of motion which is associated to this rotational symmetry is the *canonical* orbital angular momentum J_c,

$$\begin{aligned} J_c &= \boldsymbol{r}\wedge\boldsymbol{p} = \boldsymbol{r}\wedge m\boldsymbol{v} + \frac{e}{c}\boldsymbol{r}\wedge\boldsymbol{A} \\ &= \boldsymbol{r}\wedge m\boldsymbol{v} + \frac{e\Phi}{2\pi c} \\ &= J + \frac{e\Phi}{2\pi c} \end{aligned} \tag{3.1.7}$$

where J is the gauge invariant *kinetic* angular momentum. As clearly stated in (Jackiw and Redlich 1983), the conserved canonical angular momentum J_c has a conventional spectrum: Its eigenvalues are always integers in units of $\hbar$. This is true despite the fact that the algebra of the two-dimensional rotations is abelian and in principle an arbitrary constant could be added to the angular momentum operator, thus obtaining arbitrary eigenvalues.

The difference between J_c and J is entirely due to the magnetic flux Φ in the solenoid. Both in the absence and in the presence of Φ, the canonical angular momentum is *always* represented by the quantum mechanical operator

$$J_c = -i\hbar\frac{\partial}{\partial\varphi} \tag{3.1.8}$$

where φ is the polar angle on the plane. When it acts on *single-valued* wavefunctions with angular dependence $\exp(im\varphi)$, it becomes simply

$$J_c = \hbar m \quad , \quad m \in \mathbb{Z} \quad . \tag{3.1.9}$$

Thus the spectrum of J_c is always the conventional one (Jackiw and Redlich 1983). Of course when $\Phi = 0$ and hence $J_c = J$, also the kinetic angular momentum has only integer eigenvalues. However when $\Phi \neq 0$, this is not true any more. In fact using (3.1.7) and (3.1.8), we see that the kinetic angular momentum operator is

$$J = J_c - \frac{e\Phi}{2\pi c} = -i\hbar\frac{\partial}{\partial\varphi} - \frac{e\Phi}{2\pi c} \tag{3.1.10}$$

which, when acting on single-valued wavefunctions with angular dependence $\exp(im\varphi)$, becomes

$$J = \hbar\left(m - \frac{e\Phi}{hc}\right) \quad , \quad m \in \mathbb{Z} \ . \tag{3.1.11}$$

Therefore the spectrum of J consists of integers shifted by $-e\Phi/hc$.

Usually in quantum mechanics we never consider the kinetic angular momentum, since the conserved quantity is the canonical one. However for the present situation, A.S Goldhaber and R. Mackenzie (Goldhaber and Mackenzie, 1988) lucidly pointed out that the integer canonical angular momentum is divided into two pieces: One which is localized near the cyon and is in general fractional, and one which is located at the spatial infinity and is also fractional. Furthermore they argued that this diffused angular momentum is irrelevant in describing local phenomena, and identified the piece localized on the cyon with the kinetic angular momentum.

We can reformulate their argument as follows. First of all let us observe that the following identity holds

$$\begin{aligned} J_c &= J + \frac{e}{c}\, \boldsymbol{r} \wedge \boldsymbol{A} \\ &= J - \frac{1}{c}\int d^2x\ \boldsymbol{x} \cdot \boldsymbol{E}(t,\boldsymbol{x})\ B(t,\boldsymbol{x}) + \frac{1}{c}\int d^2x\ \nabla \cdot [\boldsymbol{E}(t,\boldsymbol{x})\ \boldsymbol{x} \wedge \boldsymbol{A}] \ . \end{aligned} \tag{3.1.12}$$

In this formula $\boldsymbol{E}(t,\boldsymbol{x})$ represents the electric field created by the moving charge which satisfies the Gauss law $\nabla \cdot \boldsymbol{E}(t,\boldsymbol{x}) = e\,\delta^{(2)}(\boldsymbol{x} - \boldsymbol{r}(t))$ where $\boldsymbol{r}(t)$ is the particle position at time t. [6] When we substitute (3.1.3), we find that the second term in the right-hand-side of (3.1.12) identically vanishes, so that for the cyon system we are left with

$$J_c = J + \frac{1}{c}\int d^2x\ \nabla \cdot [\boldsymbol{E}(\boldsymbol{x})\ \boldsymbol{x} \wedge \boldsymbol{A}] \ . \tag{3.1.13}$$

The difference between J_c and J is just a surface term which however cannot be neglected *tout-court* – indeed as we have seen its value is $e\Phi/2\pi c$! We can understand and clarify the role of such surface term if we imagine slowly turning on the flux through the solenoid from a zero initial value to the final value Φ. When $\Phi = 0$, $J_c = J$ and both are integers. When Φ is turned on, only J_c remains constant and thus integer. Its conservation is in fact a consequence of the rotational

[6] The reader may verify that by computing the divergence inside the last integral of (3.1.12) one produces the $\frac{e}{c}\, \boldsymbol{r} \wedge \boldsymbol{A}$ term of J_c upon using the Gauss law, plus a term of the form $\int d^2x\ \boldsymbol{E}(t,\boldsymbol{x}) \cdot \nabla(\boldsymbol{x} \wedge \boldsymbol{A})$ which exactly cancels the first integral in (3.1.12).

invariance of the system, for which J_c is the conserved Noether charge. When the flux reaches its final value Φ, a piece of J_c (specifically the surface term in (3.1.13)) is radiated away to infinity by the gauge fields. For phenomena on a finite length scale, after a finite time, this "dissipated" piece of the angular momentum is no longer relevant, and only J is left on the cyon. Thus even though the total canonical angular momentum remains integer, the cyon retains just a part of it which is in general fractional.

Following (Wilczek 1982a,b, 1990), we call this (kinetic) angular momentum the *spin* of the cyon. More precisely we have

$$s = \frac{J(m=0)}{\hbar} = -\frac{e\Phi}{hc} \ . \tag{3.1.14}$$

In general s is neither integer nor half-integer.

If some connection between spin and statistics exists, we should expect that the cyon be in general an anyon. To establish their statistical properties, let us consider two identical cyons with a wavefunction $\psi(1,2)$, and assume that the magnetic flux and the electric charge are tightly bound on each particle. Now let us slowly move one cyon around the other by a full loop and neglect both charge-charge and vortex-vortex interactions. As it was discovered by Y. Aharonov and D. Bohm (Aharonov and Bohm 1959), even though the presence of a vortex does not effect the classical dynamics of a charged particle, it has profound consequences on its quantum mechanics. Specifically in our case when, say, particle 1 is moved around the vortex 2 on a closed loop Γ , the wave function acquires a phase

$$\exp\left(-\mathrm{i}\frac{e}{\hbar c}\int_\Gamma d\boldsymbol{r}\cdot\boldsymbol{A}\right) \ . \tag{3.1.15}$$

Using Stoke's theorem, (3.1.15) can be rewritten in terms of the magnetic flux, namely

$$\exp\left(-\mathrm{i}\frac{e}{\hbar c}\int_\Gamma d\boldsymbol{r}\cdot\boldsymbol{A}\right) = \exp\left(-\mathrm{i}\frac{e}{\hbar c}\int d^2r B\right) = \exp\left(-2\pi\mathrm{i}\frac{e\Phi}{hc}\right) \ . \tag{3.1.16}$$

In the two-cyon system, when the particles are rotated around each other, there are actually *two* contributions to the phase: One due to the motion of the first charged particle in the vortex field of the second, and one due to the motion of the second charged particle in the vortex field of the first. The total phase acquired by the wavefunction $\psi(1,2)$ under a full 2π rotation is then

$$\exp\left(-2\pi\mathrm{i}\frac{2e\Phi}{hc}\right) \ . \tag{3.1.17}$$

If we compare (3.1.17) with (2.1) in Chapter 2 and recall that in this case $\Delta\varphi = 2\pi$, we can easily deduce that the statistics of the cyon is

$$\nu = -\frac{2e\Phi}{hc} \ . \tag{3.1.18}$$

Moreover the spin s and the statistics ν appear to be related in the conventional way

$$\nu = 2s \quad . \tag{3.1.19}$$

Thus in general the cyon is an anyon and the standard spin-statistics connection is satisfied! In particular, a bosonic particle with a flux $\Phi = \frac{1}{2}\Phi_0$ (where $\Phi_0 = hc/e$ is the fundamental flux unit) behaves effectively as a fermion. We conclude this section by mentioning that the cyon system possesses not only spatial rotational symmetry, but also a conformal symmetry (Jackiw 1990). In fact, one can check that under the following transformations

$$\begin{aligned} \delta \boldsymbol{r} &= \boldsymbol{v} \ , \\ \delta \boldsymbol{r} &= t\boldsymbol{v} - \frac{1}{2}\boldsymbol{r} \ , \\ \delta \boldsymbol{r} &= t^2\boldsymbol{v} - t\boldsymbol{r} \ , \end{aligned} \tag{3.1.20}$$

the Lagrangian (3.1.1) changes only by a total derivative and thus (3.1.20) are symmetries of the system. The conserved Noether charges associated to (3.1.20) are respectively

$$\begin{aligned} H &= \frac{1}{2}m\boldsymbol{v}^2 \ , \\ D &= tH - \frac{1}{4}(\boldsymbol{r}\cdot\boldsymbol{v} + \boldsymbol{v}\cdot\boldsymbol{r}) \ , \\ K &= -t^2H + 2tD + \frac{1}{2}m\boldsymbol{r}^2 \ . \end{aligned} \tag{3.1.21}$$

Using the canonical commutation relations and keeping in mind the non-trivial relationship between the canonical momentum $\boldsymbol{p}$ and the velocity $\boldsymbol{v}$ (see (3.1.5)), one may show that the charges H, D and K close the following algebra

$$\begin{aligned} [D, H] &= -\mathrm{i}\hbar H \ , \\ [D, K] &= \mathrm{i}\hbar K \ , \\ [H, K] &= 2\mathrm{i}\hbar D \ . \end{aligned} \tag{3.1.22}$$

If we define

$$\begin{aligned} L_\pm &= \frac{1}{2\hbar}(K - H) \pm \frac{\mathrm{i}}{\hbar}D \ , \\ L_0 &= -\frac{1}{2\hbar}(K + H) \ , \end{aligned} \tag{3.1.23}$$

then the commutators (3.1.22) can be rewritten as

$$\begin{aligned} [L_+, L_-] &= 2L_0 \ , \\ [L_0, L_\pm] &= \mp L_\pm \ . \end{aligned} \tag{3.1.24}$$

In these we recognize the commutation relations of $SO(2,1)$ in the standard Cartan basis. We recall that $SO(2,1)$ is the finite subalgebra of the Virasoro algebra which describes the conformal transformations in two dimensions. We will see in Chapter 9 that there is a remarkable even though mysterious connection between anyons and conformal field theory.

3.2 Chern-Simons Construction of Fractional Statistics

In the previous section we have seen that a particle which carries both a charge e and a flux Φ possesses a spin s

$$s = -\frac{e\Phi}{hc} \ , \tag{3.2.1}$$

which is neither integer nor half-integer in general, and is an anyon of statistics $\nu = 2s$. In this section we are going to present a simple construction of fractional statistics which generalizes these features to a many-body system by using Chern-Simons gauge fields. The use of the Chern-Simons theory to realize fractional statistics was first suggested by F. Wilczek (Wilczek 1982a,b) and subsequently clarified by many others (see for example Arovas *et al.* 1985; Hansson *et al.* 1988). The essential feature of the Chern-Simons formulation is that fractional statistics is implemented by means of a local, long-range, abelian gauge interaction taking place in a (2+1)-dimensional space-time. In fact rather than affixing by hand an externally prescribed vector potential $\boldsymbol{A}$ to an ordinary particle, which is then transmuted into an anyon, one promotes $\boldsymbol{A}$ to be the space component of a dynamical (2+1)-dimensional $U(1)$ gauge field A_α whose action is taken to be

$$S_{\mathrm{CS}} = \int d^3x \ \mathcal{L}_{\mathrm{CS}} = \frac{\kappa}{2c} \int d^3x \ \epsilon^{\alpha\beta\gamma} A_\alpha \partial_\beta A_\gamma \tag{3.2.2}$$

where $\epsilon^{\alpha\beta\gamma}$ is the completely antisymmetric tensor density. Here and in the following we adopt the convention that Greek indices take the values 0,1,2 and are contracted with the metric $\eta_{\alpha\beta} = \mathrm{diag}(1, -1, -1)$. The space components are instead labeled by Latin indices taking the values 1,2. The three-vector $x^\alpha \equiv (ct, x^i)$ is denoted for short also as x, so that $d^3x = cdt\, d^2x$.

The expression (3.2.2) is known as Chern-Simons action (Siegel 1979; Schonfeld 1981; Jackiw and Templeton 1981; Deser *et al.* 1982a,b; Hagen 1984). It is gauge invariant even if the Lagrangian contains an undifferentiated gauge field A_α. This is so because the gauge potential is contracted with the current $\epsilon^{\alpha\beta\gamma}\partial_\beta A_\gamma = 1/2\epsilon^{\alpha\beta\gamma}F_{\beta\gamma}$ (where as usual $F_{\beta\gamma} = \partial_\beta A_\gamma - \partial_\gamma A_\beta$) which is conserved because of the Bianchi identity. In other words under gauge transformations

$$A_\alpha \longrightarrow A_\alpha + \partial_\alpha \Lambda \tag{3.2.3}$$

where Λ is a space-time dependent parameter, the Chern-Simons Lagrangian changes only by a total derivative

$$\delta\mathcal{L}_{\mathrm{CS}} = -\frac{\kappa}{2c}\,\epsilon^{\alpha\beta\gamma}\,\partial_\alpha\left[(\partial_\beta\Lambda)\,A_\gamma\right] \ , \tag{3.2.4}$$

so that the action (3.2.2) remains invariant.

The Chern-Simons term as written here is special to (2+1) dimensions, but it can be easily generalized to any odd dimensional space-time and appears in a host of different contexts: anomalies (see for example (Alvarez-Gaumé and Ginsparg 1985)), supergravity (see for example (Ceresole *et al.* 1986)), string theory (see for example (Green and Schwarz 1984)) and so on. The Chern-Simons term provides also an example of topological field theory (for a review see (Birmingham *et al.* 1991)) since even in a curved space-time, the action has the same form as in (3.2.2) without any additional metric insertions (the ϵ-symbol is indeed a good tensor density). We refer the reader to the extensive literature on these numerous applications of the Chern-Simons term, and concentrate only on its use to construct anyons.

To this end, let us couple the gauge field A_α to a matter system consisting of N non-relativistic point particles of mass m and charge e [7], whose coordinates $\boldsymbol{r}_I(t)$ serve as dynamical variables (capital Latin indices take the values $1,\ldots,N$ and label the particles). The current

$$j^\alpha(x) = \sum_{I=1}^{N} e\, v_I^\alpha(t)\ \delta^{(2)}(\boldsymbol{x} - \boldsymbol{r}_I(t)) \equiv (c\,\rho, \boldsymbol{j}) \tag{3.2.5}$$

where $v_I^\alpha(t) \equiv (c, \boldsymbol{v}_I(t))$, clearly satisfies the continuity equation

$$\partial_\alpha j^\alpha = \partial_t \rho + \nabla \cdot \boldsymbol{j} = 0 \ . \tag{3.2.6}$$

The meaning of j^α becomes more evident if we write explicitly the time and space components, namely

$$\begin{aligned} \rho(x) &= \sum_{I=1}^{N} e\ \delta^{(2)}(\boldsymbol{x} - \boldsymbol{r}_I(t)) \ , \\ \boldsymbol{j}(x) &= \sum_{I=1}^{N} e\ \boldsymbol{v}_I(t)\ \delta^{(2)}(\boldsymbol{x} - \boldsymbol{r}_I(t)) \ . \end{aligned} \tag{3.2.7}$$

We recognize in ρ and $\boldsymbol{j}$ respectively the conventional charge density and current density for N non-relativistic point particles located at $\boldsymbol{r}_I(t)$ and moving with velocities $\boldsymbol{v}_I(t) = \dot{\boldsymbol{r}}_I(t)$.

Then we couple the conserved current j^α to the gauge field A_α in the standard minimal way

$$\begin{aligned} S_{\rm int} &= -\frac{1}{c^2}\int d^3x\ j^\alpha(x) A_\alpha(x) \\ &= \frac{1}{c}\int dt \left\{ \sum_{I=1}^{N} e\ \Big[\boldsymbol{v}_I(t) \cdot \boldsymbol{A}(t, \boldsymbol{r}_I(t)) - A_0(t, \boldsymbol{r}_I(t)) \Big] \right\} \end{aligned} \tag{3.2.8}$$

(recall that $d^3x = c dt\, d^2x$), and take as kinetic term for the N particles, the non-relativistic action

[7] Notice that this charge is *not* the standard electric charge since the gauge field A_α does not have the Maxwell action of the standard electrodynamics but the unconventional Chern-Simons term (3.2.2).

$$S_{\text{matter}} = \int dt \left(\sum_{I=1}^{N} \frac{1}{2} m \boldsymbol{v}_I^2 \right) \quad . \tag{3.2.9}$$

The total action for our system is therefore

$$S = S_{\text{matter}} + S_{\text{int}} + S_{\text{CS}} = \int dt \; \mathcal{L} \tag{3.2.10}$$

where the total Lagrangian $\mathcal{L}$, spelled out in detail, is

$$\begin{aligned} \mathcal{L} = &\sum_{I=1}^{N} \left[\frac{1}{2} m \boldsymbol{v}_I^2 + \frac{e}{c} \boldsymbol{v}_I \cdot \boldsymbol{A}(t, \boldsymbol{r}_I(t)) - \frac{e}{c} A_0(t, \boldsymbol{r}_I(t)) \right] \\ &- \frac{\kappa}{2} \int d^2x \; [\boldsymbol{E}(t, \boldsymbol{x}) \wedge \boldsymbol{A}(t, \boldsymbol{x}) + A_0(t, \boldsymbol{x}) \, B(t, \boldsymbol{x})] \quad . \end{aligned} \tag{3.2.11}$$

Here we have introduced the Chern-Simons magnetic field

$$B = \nabla \wedge \boldsymbol{A} = \partial_1 A^2 - \partial_2 A^1 \equiv -F_{12} \quad , \tag{3.2.12}$$

and the Chern-Simons electric field

$$E^i = -\frac{1}{c} \partial_t A^i - \partial_i A_0 \equiv F_{0i} \quad . \tag{3.2.13}$$

(We recall that space indices are lowered and raised with the metric $\eta_{ij} = \text{diag}(-1, -1)$.)

By varying $\mathcal{L}$ with respect to $\boldsymbol{r}_I(t)$, we obtain the equations of motion for the matter variables

$$m \, \dot{v}_I^i(t) = e \left(E^i(t, \boldsymbol{r}_I(t)) + \frac{1}{c} \epsilon^{ij} v_I^j \; B(t, \boldsymbol{r}_I(t)) \right) \tag{3.2.14}$$

where $\epsilon^{12} = -\epsilon^{21} = 1$. These are the standard Lorentz force equations for particles of mass m and charge e moving in an electric field $\boldsymbol{E}$ and in a magnetic field B. By varying $\mathcal{L}$ with respect to A_α we get instead the equations of motion for the gauge fields. However, because of the peculiar nature of the Chern-Simons term which contains only first-derivatives of A_α, more than true equations of motion these are actually identities that relate the fields $\boldsymbol{E}$ and B to the matter current $\boldsymbol{j}$ and ρ; indeed one gets

$$j^\alpha = \frac{\kappa \, c}{2} \epsilon^{\alpha\beta\gamma} F_{\beta\gamma} \quad , \tag{3.2.15}$$

or in components

$$E^i = \frac{1}{\kappa \, c} \epsilon^{ij} j^j \quad , \tag{3.2.16a}$$

$$B = -\frac{1}{\kappa} \rho \quad . \tag{3.2.16b}$$

These equations are very important. First of all, they tell us that given a matter configuration (that is given ρ and $\boldsymbol{j}$), the field strengths of the Chern-Simons fields are not arbitrary but are uniquely prescribed according to (3.2.16). This is very different from the cyon system of Section 2 where the magnetic field B was an externally given field and could be *chosen* to be that of a vortex. Secondly, if we use (3.2.16) for a single particle ($N = 1$ and subscript I suppressed), it is easy to prove that

$$\begin{aligned} E^i(x) + \frac{1}{c}\epsilon^{ij}v^j(t)\ B(x) &= \frac{1}{\kappa c}\epsilon^{ij}\left(j^j(x) - v^j(t)\ \rho(x)\right) \\ &= \frac{e}{\kappa c}\epsilon^{ij}\left(v^j(t) - v^j(t)\right)\ \delta^{(2)}(\boldsymbol{x} - \boldsymbol{r}(t)) = 0 \ . \end{aligned} \tag{3.2.17}$$

In going from the first to the second line, we have used the appropriate explicit expressions (3.2.7). Thus, a particle moving in a Chern-Simons background does not experience any self-interaction since the Lorentz force is vanishing. However, due to the non-trivial relationship between the velocity $\boldsymbol{v}$ and the canonical momentum $\boldsymbol{p}$, *i.e.*

$$\boldsymbol{p} = m\boldsymbol{v} + \frac{e}{c}\boldsymbol{A}(\boldsymbol{r}) \ , \tag{3.2.18}$$

the presence of the Chern-Simons electric and magnetic fields is not at all immaterial. For an N particle system, the equations of motion are explicitly given by

$$m\dot{v}_I^i = \frac{e^2}{\kappa c}\sum_{J=1}^{N}\epsilon^{ij}\left(v_J^j(t) - v_I^j(t)\right)\delta^{(2)}(\boldsymbol{r}_I(t) - \boldsymbol{r}_J(t)) \ . \tag{3.2.19}$$

Notice that for $J = I$ the coefficient of the ill-defined $\delta^{(2)}(\boldsymbol{r}_I(t) - \boldsymbol{r}_I(t))$ on the r.h.s. of (3.2.19), identically vanishes, and hence no ambiguities and no self-interactions are present in the system (Jackiw 1990).

The vanishing of the Lorentz force does not come as a total surprise (De Sousa Gerbert 1990). In fact as is well-known, (3.2.14) can also be written in covariant notation as

$$\partial_\alpha T^{\alpha\beta}_{\text{matter}} = -\frac{1}{\kappa c}f^\beta \tag{3.2.20}$$

where $T^{\alpha\beta}_{\text{matter}}$ is the energy-momentum tensor of the matter system and f^β is the force density given by

$$f^\beta = F^{\beta\gamma}j_\gamma = \left(E^i j^i, E^i j^0 + \epsilon_{ik}j^k B\right) \ . \tag{3.2.21}$$

Using (3.2.16), we easily see that $f^\beta = 0$ identically, and thus $T^{\alpha\beta}_{\text{matter}}$ is conserved. The vanishing of f^β and hence the conservation of $T^{\alpha\beta}_{\text{matter}}$ are necessary because the Chern-Simons action, being topological and metric independent, does not contribute to the energy momentum tensor. So $T^{\alpha\beta}_{\text{matter}}$ is the *full* energy momentum tensor of the whole system which better be conserved!

The most remarkable consequence of the Chern-Simons interaction is however the relation (3.2.16b) between the magnetic field B and the charge density ρ which we rewrite here for convenience

$$B = -\frac{1}{\kappa}\rho \ . \tag{3.2.22}$$

If we integrate this equation over a small two-dimensional disc C_I which includes only the I-th particle, the left-hand-side yields the magnetic flux attached to that particle while the right-hand-side, upon using the explicit expression (3.2.7), gives its charge, namely

$$\Phi_I = \int_{C_I} d^2x\; B = -\frac{e}{\kappa}\int_{C_I} d^2x \sum_{J=1}^{N} \delta^{(2)}\left(\boldsymbol{x} - \boldsymbol{r}_J(t)\right) = -\frac{e}{\kappa} \; . \tag{3.2.23}$$

Therefore, whenever a particle possesses a charge e, it necessarily possesses also a flux $\Phi = -e/\kappa$. The Chern-Simons dynamics automatically binds charge and flux to the particles, and thus realizes in a natural way what had to be imposed as an external requirement in the cyon system. We can picture the situation as follows: each particle possesses both a charge and a flux, and moves in the background fields created by the other particles; the Chern-Simons electric and magnetic fields conspire to cancel all self-interactions and the resulting Lorentz force is zero. However, since a charge interacts with all other fluxes and vice versa, we expect interesting quantum effects from the Aharonov-Bohm mechanism.

The Hamiltonian arising via canonical procedure from the total Lagrangian (3.2.11), is

$$H = \sum_{I=1}^{N}\left(\frac{1}{2}m v_I^2\right) + \int d^2x\; A_0(x)\left(\kappa B(x) + \rho(x)\right) \; . \tag{3.2.24}$$

If $\rho = -\kappa\, B$ is imposed as a constraint, A_0 may be set equal to zero by choosing the Weyl gauge, and hence H becomes numerically equal to the Hamiltonian of N non-interacting particles

$$H = \sum_{I=1}^{N}\left(\frac{1}{2}m v_I^2\right) \; . \tag{3.2.25}$$

As in the cyon system, the non-trivial dynamics resides entirely in the relation between the canonical and the kinetic momenta. The condition $A_0 = 0$ does not fix the gauge completely since we are still free to perform any time-independent gauge transformation without leaving the Weyl gauge. This residual freedom is removed by imposing the subsidiary condition

$$\partial_i A^i = 0 \; . \tag{3.2.26}$$

Now the constraint $\rho = -\kappa\, B = \kappa\epsilon^{ij}\partial_i A_j$ with ρ given in (3.2.7), can be unambiguously solved and one finds

$$A_I^i(\boldsymbol{r}_1, \ldots \boldsymbol{r}_N) = \frac{e}{2\pi\kappa}\sum_{J\neq I}\epsilon^{ij}\frac{r_I^j - r_J^j}{|\boldsymbol{r}_I - \boldsymbol{r}_J|^2} \; . \tag{3.2.27}$$

Thus, by eliminating the Chern-Simons gauge field A_α through its field equations (or more properly through the constraints (3.2.16) after gauge fixing), we end up with a system of particles moving in the effective non-local vector potential (3.2.27). The Hamiltonian (3.2.25) can then be rewritten as follows

$$H' = \sum_{I=1}^{N} \frac{1}{2m} \left(\boldsymbol{p}_I - \frac{e}{c} \boldsymbol{A}_I(\boldsymbol{r}_1, ...\boldsymbol{r}_N) \right)^2 \quad , \tag{3.2.28}$$

where the prime on H denotes that we used the solution given in (3.2.27). We remark once again that this effective vector potential is non-local since it depends on the positions of all particles, and in particular when $N = 1$ it vanishes. If we compute the magnetic field associated to $\boldsymbol{A}_I(\boldsymbol{r}_1, ...\boldsymbol{r}_N)$, we get

$$B_I = \epsilon^{ij} \frac{\partial}{\partial r_I^i} A_I^j(\boldsymbol{r}_1, ...\boldsymbol{r}_N) = -\frac{e}{\kappa} \sum_{J \neq I} \delta^{(2)}(\boldsymbol{r}_I - \boldsymbol{r}_J) \quad . \tag{3.2.29}$$

Each particle sees the $(N-1)$ others as vortices carrying a flux $\Phi = -e/\kappa$. This is in agreement with our previous derivation (3.2.23) based on the use of the Chern-Simons constraints.

The non-local potential $\boldsymbol{A}_I(\boldsymbol{r}_1, ...\boldsymbol{r}_N)$ seems badly divergent at first sight. However this is not the case because the would-be singularities occur only at coincident points ($\boldsymbol{r}_I = \boldsymbol{r}_J$ for some I and J) which do not belong to the configuration space of the system (see (2.11)).

It is interesting to observe that the solution (3.2.27) can be written also as

$$\begin{aligned} A_I^i(\boldsymbol{r}_1, ...\boldsymbol{r}_N) &= \frac{e}{2\pi\kappa} \epsilon^{ij} \frac{\partial}{\partial r_I^j} \sum_{J \neq I} \ln |\boldsymbol{r}_I - \boldsymbol{r}_J| \\ &= -\frac{e}{2\pi\kappa} \frac{\partial}{\partial r_I^i} \sum_{J \neq I} \varphi_{IJ} \end{aligned} \tag{3.2.30}$$

where φ_{IJ} is the winding angle of particle J with respect to particle I, such that

$$\varphi_{IJ} = \tan^{-1} \left(\frac{x_I^2 - x_J^2}{x_I^1 - x_J^1} \right) \tag{3.2.31}$$

(see (2.22)). The Lagrangian corresponding to (3.2.28) is then

$$\begin{aligned} \mathcal{L}' &= \sum_{I=1}^{N} \left(\frac{1}{2} m v_I^2 + \frac{e}{c} \, \boldsymbol{v}_I \cdot \boldsymbol{A}_I(\boldsymbol{r}_1, ...\boldsymbol{r}_N) \right) \\ &= \sum_{I=1}^{N} \left(\frac{1}{2} m v_I^2 \right) - \frac{e^2}{2\pi c\kappa} \sum_{I=1}^{N} \sum_{J \neq I} v_I^i \frac{\partial}{\partial r_I^i} \varphi_{IJ} \\ &= \sum_{I=1}^{N} \left(\frac{1}{2} m v_I^2 \right) - \frac{e^2}{2\pi c\kappa} \sum_{I<J} \left(v_I^i - v_J^i \right) \frac{\partial}{\partial r_I^i} \varphi_{IJ} \end{aligned} \tag{3.2.32}$$

where in deriving the last term we used the property

$$\frac{\partial}{\partial r_J^i} \varphi_{IJ} = -\frac{\partial}{\partial r_I^i} \varphi_{IJ} \quad . \tag{3.2.33}$$

If we now observe that

$$\frac{d}{dt}\varphi_{IJ} = \left(v_I^i \frac{\partial}{\partial r_I^i} + v_J^i \frac{\partial}{\partial r_J^i}\right) \varphi_{IJ} = \left(v_I^i - v_J^i\right) \frac{\partial}{\partial r_I^i}\varphi_{IJ} \quad , \tag{3.2.34}$$

$\mathcal{L}'$ becomes simply

$$\mathcal{L}' = \sum_{I=1}^{N} \left(\frac{1}{2} m v_I^2\right) - \frac{e^2}{2\pi c\kappa} \left(\sum_{I<J} \frac{d}{dt}\varphi_{IJ}\right) \quad . \tag{3.2.35}$$

We managed to cast our Lagrangian in the form of (2.30) and so we immediately deduce that our particle are generically anyons of statistics

$$\nu = \frac{e^2}{2\pi\, \hbar\, c\, \kappa} = -\frac{e\,\Phi}{2\pi\, \hbar\, c} = \frac{\kappa\, \Phi^2}{2\pi\, \hbar\, c} \tag{3.2.36}$$

where the second two equalities follow from the charge-flux relation (3.2.23). Thus we can turn bosons into fermions, fermions into bosons or in general anyons into other anyons by suitably adjusting the coupling constant κ of the Chern-Simons term. Notice that the statistics ν in (3.2.36) is one half of the one computed in the previous section for the cyon system (Hansson *et al.* 1988; Wen and Zee 1989a). It is now interesting to see how this factor of 1/2 comes about. To this end let us consider again the total Lagrangian (3.2.11) and observe that under a gauge transformation with parameter Λ

$$A_\alpha \longrightarrow A_\alpha + \partial \Lambda \quad , \tag{3.2.37}$$

its variation is

$$\delta\mathcal{L} = \int d^2x \left(-\frac{1}{c} j^\alpha + \frac{\kappa}{4} \epsilon^{\alpha\beta\gamma} F_{\beta\gamma}\right) \partial_\alpha \Lambda \tag{3.2.38}$$

where j^α is the matter current (3.2.5) and the $\epsilon\, F$ term originates from the Chern-Simons action (see (3.2.4)). The Noether current associated to the gauge transformation (3.2.37) is

$$J^\alpha = \frac{1}{c} j^\alpha - \frac{\kappa}{4} \epsilon^{\alpha\beta\gamma} F_{\beta\gamma} \quad . \tag{3.2.39}$$

Both terms in (3.2.39) are clearly conserved (because of the continuity equation on the matter current and the Bianchi identity on the Chern-Simons current) and hence the total action $S = \int dt\, \mathcal{L}$ is invariant against $U(1)$ gauge transformations. Using the Chern-Simons constraint (3.2.22), we can rewrite the Noether current (3.2.39) as

$$J^\alpha = \frac{1}{2c} j^\alpha \quad . \tag{3.2.40}$$

If we integrate its time component over a small disc C_I containing only the I-th particle, we deduce that its conserved $U(1)$ Noether charge is

$$q_I = \int_{C_I} d^2x\, J^0 = \frac{1}{2} \int_{C_I} d^2x\, \rho = \frac{1}{2} e \quad . \tag{3.2.41}$$

The charge associated to the $U(1)$ Chern-Simons field is *not* e but 1/2 e! The use of the Chern-Simons action action modifies the standard electrodynamic relation by introducing a factor of 1/2. In fact, if we had considered the ordinary Maxwell term instead of the Chern-Simons one, the $U(1)$ Noether current would have been

$$J'^{\alpha} = \frac{1}{c} j^{\alpha} \quad , \tag{3.2.42}$$

and correspondingly the $U(1)$ Noether charge (that is the standard electric charge) would have been

$$q'_I = \int_{C_I} d^2x \; J'^0 = \int_{C_I} d^2x \; \rho = e \quad . \tag{3.2.43}$$

The difference originates from the fact that the Maxwell Lagrangian is invariant under gauge transformations, whilst the Chern-Simons Lagrangian varies into a total derivative.

Let us now compute the statistical phase induced by the Aharonov-Bohm effect on our particles. Let us consider a two-body wave function $\psi(1,2)$ and suppose that particle 2 is moved on a full loop around particle 1. Under such operation, due to the interaction between the charge of particle 2 and the flux of particle 1, the wave function acquires a phase

$$\exp\left(-\mathrm{i}\frac{q}{\hbar c}\int_{\Gamma} d\boldsymbol{r}\cdot\boldsymbol{A}\right) = \exp\left(-\mathrm{i}\frac{q}{\hbar c}\int d^2r B\right) = \exp\left(-\mathrm{i}\frac{q\Phi}{\hbar c}\right) \tag{3.2.44}$$

where q is the Noether charge (3.2.41) (we suppressed the index on q because the particles are assumed to have the same charge). To get the complete phase, we must multiply the exponent of (3.2.44) by two to count also the interaction between the charge of particle 1 and the flux of particle 2. In conclusion we obtain that the total phase acquired by $\psi(1,2)$ is

$$\exp\left(-\mathrm{i}\frac{2q\Phi}{\hbar c}\right) = \exp\left(-2\pi\mathrm{i}\frac{2q\Phi}{hc}\right) \quad . \tag{3.2.45}$$

Using (2.1) and remembering that $\Delta\varphi = 2\pi$ for a full loop Γ, we deduce that

$$\nu = -\frac{2q\Phi}{hc} = -\frac{e\Phi}{2\pi\hbar c} \quad , \tag{3.2.46}$$

in perfect agreement with our previous calculation (see (3.2.36)).

In particular a bosonic particle with a flux $\Phi = 1/2\ \tilde{\Phi}_0$ (where $\tilde{\Phi}_0 = hc/q = 2hc/e$ is the fundamental flux unit of the Chern-Simons theory) is a fermion. Notice that the Chern-Simons flux unit $\tilde{\Phi}_0 = hc/q$ is twice the ordinary electrodynamic flux unit $\Phi_0 = hc/e$ because of (3.2.41). However both in the cyon system and in the Chern-simons theory, one can say that a boson is converted into a fermion by affixing 1/2 unit of flux!

The system described by the total Lagrangian (3.2.11) is invariant under rotations of the spatial coordinates

$$\delta r_I^i = -\epsilon^{ij} r_I^j \quad , \tag{3.2.47}$$

and of the gauge potentials

$$\begin{aligned}\delta A_0(t,\boldsymbol{x}) &= -\epsilon^{jk}x^j\partial_k A_0(t,\boldsymbol{x}) \ , \\ \delta A_i(t,\boldsymbol{x}) &= -\epsilon^{jk}x^j\partial_k A_i(t,\boldsymbol{x}) - \epsilon^{ij}A_j(t,\boldsymbol{x}) \ .\end{aligned} \tag{3.2.48}$$

The corresponding Noether charge is the total angular momentum

$$J_{\text{tot}} = \sum_{I=1}^{N} \boldsymbol{r}_I \wedge \boldsymbol{p}_I + \frac{\kappa}{c}\int d^2x\,(\boldsymbol{x}\wedge \boldsymbol{A}(x))\,B(x) \ . \tag{3.2.49}$$

The first term in (3.2.49) is the canonical angular momentum for the N particle system, whilst the second term is the explicit contribution of the Chern-Simons fields. Using the constraint (3.2.22), we can rewrite (3.2.49) as

$$\begin{aligned}J_{\text{tot}} &= \sum_{I=1}^{N} \boldsymbol{r}_I \wedge \boldsymbol{p}_I - \frac{e}{c}\sum_{I=1}^{N} \boldsymbol{r}_I \wedge \boldsymbol{A}(\boldsymbol{r}_I) \\ &= \sum_{I=1}^{N} \boldsymbol{r}_I \wedge m\boldsymbol{v}_I \equiv J \ .\end{aligned} \tag{3.2.50}$$

Thus J_{tot} coincides with the total kinetic angular momentum of the N-particle system. Furthermore, we observe that, using (3.2.7), J_{tot} can be written in terms of the current density as follows

$$J_{\text{tot}} = J = \frac{m}{e}\int d^2x\ \boldsymbol{x}\wedge \boldsymbol{j}(x) \ . \tag{3.2.51}$$

Contrary to the cyon system where we had to argue that the relevant part of the angular momentum for long time-scale phenomena was the gauge invariant kinetic piece, here in the Chern-Simons theory the conventional derivation shows directly that J *is* the conserved Noether charge for rotations. There is no contradiction in this, since, as we have already remarked, for one particle the Chern-Simons gauge fields are irrelevant (the effective vector potential (3.2.27) vanishes for $N = 1$). In this formulation in fact, the statistical gauge interaction only occurs among couples of different particles.

From (3.2.50) we expect that the spectrum of $J_{\text{tot}} = J$ consists of in general non-integer eigenvalues. To see this explicitly, we can use the relation between $\boldsymbol{p}_I$ and $m\boldsymbol{v}_I$ in the effective theory described by the Hamiltonian H' given in (3.2.28), and write

$$\begin{aligned}J = \sum_{I=1}^{N} \boldsymbol{r}_I \wedge m\boldsymbol{v}_I &= \sum_{I=1}^{N} \boldsymbol{r}_I \wedge \left(\boldsymbol{p}_I - \frac{e}{c}\ \boldsymbol{A}_I(\boldsymbol{r}_1,...\boldsymbol{r}_N)\right) \\ &= J_c - \frac{e}{c}\sum_{I=1}^{N} \epsilon^{ij} r_I^i A_I^j(\boldsymbol{r}_1,...\boldsymbol{r}_N) \ .\end{aligned} \tag{3.2.52}$$

Using the explicit expression of the vector potential given in (3.2.27) and keeping in mind that J_c has only integer eigenvalues, the spectrum of J is

$$\text{integer} + \left\{ \nu \frac{1}{2} N(N-1) \right\}_{\text{P}} \equiv \text{integer} + \ell_\nu \tag{3.2.53}$$

where the statistics $\nu = -e\Phi/2\pi\hbar c$ as in (3.2.46) is defined modulo 2, and the symbol $\{x\}_{\text{P}}$ denotes the periodic extension of x over a period of 2 (see Fig. 3.1) (Dunne *et al.* 1992b).

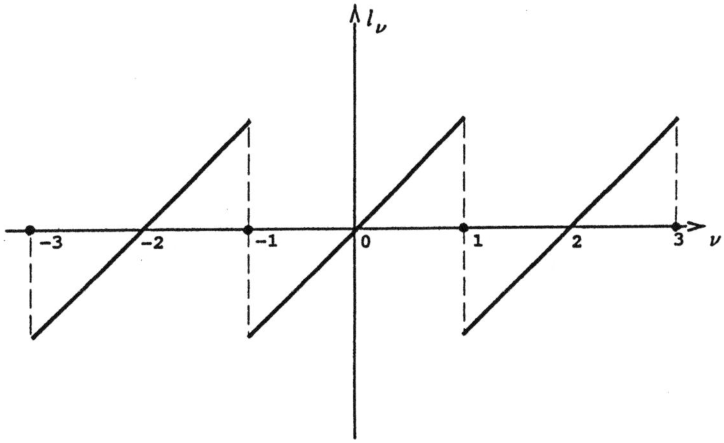

Fig. 3.1. The fractional orbital angular momentum ℓ_ν as function of the statistics ν.

In general the eigenvalues of the angular momentum are neither integers nor half-integers (in units of $\hbar$). Notice that for $\nu = 1$, $\ell_{\nu=1} = 0$. In fact for any N, $N(N-1)/2$ is always an integer and hence it can be reabsorbed into a suitable shift of J_{c}. Furthermore, for $N = 1$ $\ell_\nu = 0$ for any ν. Thus ℓ_ν has nothing to do with the "spin" of the anyons; on the contrary it is really a genuine fractional *orbital* angular momentum. The non-relativistic Chern-Simons theory, as we have presented it here, provides a framework within which the spin is an absolutely conserved quantity, so that it is perfectly consistent to ignore it, or simply set it to zero. Therefore we can say that the non-relativistic Chern-Simons theory describes spinless particles with fractional statistics. (We should not be surprised at this fact; after all it is customary to consider spinless fermions in non-relativistic quantum mechanics!) Since there is no spin for our particles, of course there is no spin-statistics connection. In Chapter 5 we will see that there is a natural way to incorporate the spin in the theory using the second-quantized formalism. Then, we will find that the usual spin-statistics relation is satisfied.

We conclude our presentation of the Chern-Simons theory by mentioning that the system described by (3.2.11) is invariant also under the following conformal transformations (Jackiw 1990)

$$\begin{aligned} &\delta \boldsymbol{r}_I = \boldsymbol{v}_I \ , \\ &\delta A_\alpha = c F_{0\alpha} + \partial_\alpha (c A_0) \ ; \end{aligned} \tag{3.2.54a}$$

$$\begin{aligned} &\delta \boldsymbol{r}_I = t\, \boldsymbol{v}_I - \frac{1}{2} \boldsymbol{r}_I \ , \\ &\delta A_\alpha = ct\, F_{0\alpha} + \partial_\alpha (ct\, A_0) - \frac{1}{2} x^i F_{i\alpha} - \frac{1}{2} \partial_\alpha (x^i A_i) \ ; \end{aligned} \tag{3.2.54b}$$

$$\begin{aligned} \delta \boldsymbol{r}_I &= t^2\, \boldsymbol{v}_I - t\, \boldsymbol{r}_I \ , \\ \delta A_\alpha &= c t^2\, F_{0\alpha} + \partial_\alpha (c t^2\, A_0) - t\, x^i\, F_{i\alpha} - \partial_\alpha (t\, x^i\, A_i) \ . \end{aligned} \tag{3.2.54c}$$

To be more precise, the Lagrangian (3.2.11) is not invariant under (3.2.54) but its variation is a total derivative so that these transformations are true symmetries of the system. The conserved Noether charges associated to (3.2.54) are respectively

$$\begin{aligned} H &= \sum_{I=1}^{N} \left(\frac{1}{2} m \boldsymbol{v}_I^2 \right) + \int d^2x\; A_0(x) \left(\kappa B(x) + \rho(x) \right) \ , \\ D &= t\, H - \frac{1}{4} m \sum_{I=1}^{N} \left(\boldsymbol{r}_I \cdot \boldsymbol{v}_I + \boldsymbol{v}_I \cdot \boldsymbol{r}_I \right) \\ &\quad + \int d^2x \left(t\, A_0(x) - \frac{1}{2c} \boldsymbol{x} \cdot \boldsymbol{A}(x) \right) \left(\kappa B(x) + \rho(x) \right) \ , \\ K &= -\, t^2\, H + 2t\, D + \sum_{I=1}^{N} \left(\frac{1}{2} m \boldsymbol{r}_I^2 \right) \\ &\quad + \int d^2x \left(t^2\, A_0(x) - \frac{t}{c} \boldsymbol{x} \cdot \boldsymbol{A}(x) \right) \left(\kappa B(x) + \rho(x) \right) \ . \end{aligned} \tag{3.2.55}$$

Upon using the constraint (3.2.22), the gauge field contributions to H, D and K disappear and one is left with the straightforward N-particle generalization of the conformal generators given in (3.1.21) for the cyon system. From the canonical commutation relations it follows that

$$\begin{aligned} [L_+, L_-] &= 2 L_0 \ , \\ [L_0, L_\pm] &= \mp L_\pm \ . \end{aligned} \tag{3.2.56}$$

where, as in (3.1.23),

$$\begin{aligned} L_\pm &= \frac{1}{2\hbar}(K - H) \pm \frac{\mathrm{i}}{\hbar} D \ , \\ L_0 &= -\frac{1}{2\hbar}(K + H) \ . \end{aligned}$$

This is the algebra of the conformal group $SO(2,1)$ in two dimensions, which is a symmetry group of the Chern-Simons theory.

4. Fractional Statistics in the Anyon Gauge

There is a very simple way to check if a many-body wavefunction describes bosons or fermions: If under an exchange of any two particles ψ acquires a plus sign, it describes bosons; on the other hand if it acquires a minus sign, it describes fermions. The statistics is therefore encoded in the boundary conditions of ψ. In the previous chapter we have seen that fractional statistics could be realized by means of an effective non-local interaction among particle of ordinary statistics. Now we are going to demonstrate that there is an equivalent description of fractional statistics where complicated boundary conditions replace the effective interaction (Wu 1984b). The two descriptions are related to each other by a (singular) gauge transformation and thus can be used interchangeably.

Let us consider again the Hamiltonian (3.2.28), which we rewrite here for convenience

$$H' = \sum_{I=1}^{N} \frac{1}{2m} \left(\boldsymbol{p}_I - \frac{e}{c}\, \boldsymbol{A}_I(\boldsymbol{r}_1, ...\boldsymbol{r}_N) \right)^2 \tag{4.1}$$

where

$$\begin{aligned} A_I^i(\boldsymbol{r}_1, ...\boldsymbol{r}_N) &= \frac{e}{2\pi\kappa} \sum_{J\neq I} \epsilon^{ij} \frac{r_I^j - r_J^j}{|\boldsymbol{r}_I - \boldsymbol{r}_J|^2} \\ &= -\frac{e}{2\pi\kappa} \frac{\partial}{\partial r_I^i} \sum_{J\neq I} \varphi_{IJ} \ , \\ \varphi_{IJ} &= \tan^{-1}\left(\frac{x_I^2 - x_J^2}{x_I^1 - x_J^1} \right) \ . \end{aligned} \tag{4.2}$$

This is the first-quantized Hamiltonian for a system of N anyons of statistics

$$\nu = \frac{e^2}{2\pi\kappa c} \ , \tag{4.3}$$

which are described by a single-valued, symmetric wavefunction ψ such that

$$H' \psi = E \psi \ . \tag{4.4}$$

We want to emphasize that ψ is single-valued and "bosonic" according to our definition, because

$$\psi(\boldsymbol{r}_1, \cdots, \boldsymbol{r}_J, \cdots, \boldsymbol{r}_I, \cdots, \boldsymbol{r}_N) = \psi(\boldsymbol{r}_1, \cdots, \boldsymbol{r}_I, \cdots, \boldsymbol{r}_J, \cdots, \boldsymbol{r}_N) \ . \tag{4.5}$$

The non-standard statistics ν is explicitly reproduced by the non-local potential $\boldsymbol{A}_I$. However, from (4.2) we see that $\boldsymbol{A}_I$ is a pure gauge and can be removed by a gauge transformation. The transformed wavefunction is

$$\tilde{\psi} = \mathrm{e}^{\mathrm{i}\nu \sum_{I<J} \varphi_{IJ}} \psi \ , \tag{4.6}$$

and the transformed Hamiltonian is

$$\begin{aligned} \tilde{H} &= \left(\mathrm{e}^{\mathrm{i}\nu \sum_{I<J} \varphi_{IJ}}\right) H \left(\mathrm{e}^{-\mathrm{i}\nu \sum_{I<J} \varphi_{IJ}}\right) \\ &= \sum_{I=1}^{N} \frac{1}{2m} \boldsymbol{p}_I^2 \ , \end{aligned} \tag{4.7}$$

so that the eigenvalue problem (4.4) becomes

$$\tilde{H}\,\tilde{\psi} = E\,\tilde{\psi} \ . \tag{4.8}$$

$\tilde{H}$ is a free Hamiltonian, but the new wavefunctions are multi-valued and satisfy the "twisted" boundary conditions

$$\tilde{\psi}(\boldsymbol{r}_1, \cdots, \boldsymbol{r}_J, \cdots, \boldsymbol{r}_I, \cdots, \boldsymbol{r}_N) = \mathrm{e}^{\mathrm{i}\pi\nu}\, \tilde{\psi}(\boldsymbol{r}_1, \cdots, \boldsymbol{r}_I, \cdots, \boldsymbol{r}_J, \cdots, \boldsymbol{r}_N) \ . \tag{4.9}$$

This equation, which follows from (4.6) and $\varphi_{JI} = \varphi_{IJ} + \pi$, explicitly shows that $\tilde{\psi}$ carries an abelian representation of the braid group B_N. Extending our initial definition, we may therefore say that $\tilde{\psi}$ is an "anyonic" wavefunction. It is customary call this the anyon gauge description of fractional statistics.

This result can be retrieved in another way which actually shows how general the anyon gauge description is, because it does not rely on the specific features of the model. We saw in Chapter 3 that the Lagrangian associated to H' is

$$\mathcal{L}' = \sum_{I=1}^{N} \left(\frac{1}{2} m v_I^2\right) - \nu \left(\sum_{I<J} \frac{d}{dt} \varphi_{IJ}\right) \ . \tag{4.10}$$

This is also the general form of the Lagrangian describing anyons of statistics ν as we discussed in Chapter 2. The kernel $K(q', t'; q, t)$, which is built out of $\mathcal{L}'$, is (see (2.29)) [8]

$$K(q', t'; q, t) = \int \mathcal{D}\tilde{q}\ e^{\frac{i}{\hbar} \int_t^{t'} d\tau \left\{ \mathcal{L} - \hbar\nu \sum_{I<J} \frac{d}{dt} \varphi_{IJ} \right\}} \ , \tag{4.11}$$

and propagates the single-valued wavefunctions according to

[8] In Chapter 2 we considered for simplicity only closed paths in the configuration space, but the construction can be clearly generalized to arbitrary open paths.

$$\psi(q',t') = \int dq \ \langle q',t'|q,t\rangle \ \langle q,t|\psi\rangle$$
$$= \int dq \ K(q',t';q,t) \ \psi(q,t) \quad . \tag{4.12}$$

Since the "statistical" interaction is a total time-derivative, (4.11) can be rewritten as (Wu 1984b)

$$K(q',t';q,t) = \sum_{n_{IJ}=-\infty}^{\infty} \mathrm{e}^{-\mathrm{i}\nu\sum\limits_{I<J}[\varphi_{IJ}(t')+2\pi n_{IJ}]} K_0^{\{n\}}(q',t';q,t) \ \mathrm{e}^{\mathrm{i}\nu\sum\limits_{I<J}\varphi_{IJ}(t)} \tag{4.13}$$

The sums over the integers n_{IJ} are a consequence of the multi-valuedness of the angles φ_{IJ}, and each contribution in these sums corresponds to paths with winding numbers n_{IJ} in the configuration space. Finally, $K_0^{\{n\}}$ is computed only from paths with fixed winding numbers, using the free Lagrangian $\mathcal{L}$.

Given (4.13) and (4.12), it is natural to redefine the wavefunction and introduce

$$\tilde{\psi} = \mathrm{e}^{\mathrm{i}\nu\sum\limits_{I<J}\varphi_{IJ}} \ \psi \quad . \tag{4.14}$$

Furthermore, if we understand the angles φ_{IJ} as defined modulo 2π so that the different windings n_{IJ} are automatically taken into account, we can simplify the notation and rewrite (4.12) as follows

$$\tilde{\psi}(q',t') = \int dq' \ K_0(q',t';q,t) \, \tilde{\psi}(q,t) \quad . \tag{4.15}$$

The wavefunction $\tilde{\psi}$ propagates with the free kernel K_0, but of course it is multi-valued. The transformation from ψ to $\tilde{\psi}$ in (4.14) and that from K to K_0 in (4.13) are obviously the same as those in (4.6) and (4.7) which allowed us to eliminate the effective non-local long-range interaction of the Chern-Simons construction, and can be performed in general whenever the Lagrangian has the form (4.10).

The wavefunction $\tilde{\psi}$ in (4.14) can also be written in a different way which is more useful for applications. Let us introduce for each particle the complex coordinates z_I and $\bar{z}_I$ in the following way

$$z_I = x_I^1 + \mathrm{i}\, x_I^2 \quad ,$$
$$\bar{z}_I = x_I^1 - \mathrm{i}\, x_I^2 \quad , \tag{4.16}$$

where (x_I^1, x_I^2) are the Cartesian coordinates of the I-th particle. It is then easy to see that

$$z_{IJ} \equiv z_I - z_J = |z_I - z_J| \ \mathrm{e}^{\mathrm{i}\varphi_{IJ}} \quad ,$$
$$\bar{z}_{IJ} \equiv \bar{z}_I - \bar{z}_J = |\bar{z}_I - \bar{z}_J| \ \mathrm{e}^{-\mathrm{i}\varphi_{IJ}} \quad . \tag{4.17}$$

Therefore, (4.14) can be written as

$$\begin{aligned}\tilde{\psi} &= \prod_{I<J} (z_{IJ})^{\nu} \left(\frac{\psi}{\prod\limits_{I<J} |z_{IJ}|^{\nu}} \right) \\ &= \prod_{I<J} (z_{IJ})^{\nu}\ P(z_1, \ldots, z_N; \bar{z}_1, \ldots, \bar{z}_N) \ ,\end{aligned} \tag{4.18}$$

or equivalently as

$$\begin{aligned}\tilde{\psi} &= \prod_{I<J} (\bar{z}_{IJ})^{-\nu} \left(\frac{\psi}{\prod\limits_{I<J} |z_{IJ}|^{-\nu}} \right) \\ &= \prod_{I<J} (\bar{z}_{IJ})^{-\nu}\ P'(z_1, \ldots, z_N; \bar{z}_1, \ldots, \bar{z}_N) \ .\end{aligned} \tag{4.19}$$

In these equations P and P' denote *single-valued* functions of the particle positions.

The generic form of the wavefunction in the anyon gauge is the one presented in (4.18) and (4.19), namely a prefactor (either $\prod\limits_{I<J} (z_{IJ})^{\nu}$ or $\prod\limits_{I<J} (\bar{z}_{IJ})^{-\nu}$) times a single-valued function.

It is interesting to observe that these prefactors appear as the obvious generalization of the Jastrow-type prefactors that are usually considered for fermions and, as we will see in Chapter 9, they can be obtained as expectation values of vertex operators in conformal field theory (Fubini 1991; Fubini and Lütken 1991; Stone 1991a; Dunne *et al.* 1991a,b; Moore and Read 1991; Cristofano *et al.* 1991a,b,c). Furthermore we point out that the Laughlin wavefunction for the ground state of the fractional quantum Hall effect (Laughlin 1983) has precisely the form (4.18) with ν an odd integer, and that the wavefunction for the excitations above the ground state has the same form as in (4.18) with ν a rational number. We will return to these points later (see Chapter 8) after introducing the second-quantized formalism for anyons, and solving the eigenvalue problem (4.4) or (4.8) in some simple models.

5. Non-relativistic Chern-Simons Field Theory

In the previous two chapters we have shown that a system described by the Hamiltonian

$$H' = \sum_{I=1}^{N} \frac{1}{2m} \left(\boldsymbol{p}_I - \frac{e}{c} \boldsymbol{A}_I(\boldsymbol{r}_1, ... \boldsymbol{r}_N) \right)^2 \quad , \tag{5.1}$$

where

$$A_I^i(\boldsymbol{r}_1, ... \boldsymbol{r}_N) = \frac{e}{2\pi\kappa} \sum_{J \neq I} \epsilon^{ij} \frac{r_I^j - r_J^j}{|\boldsymbol{r}_I - \boldsymbol{r}_J|^2} \quad , \tag{5.2}$$

and by single-valued wavefunctions ψ such that

$$\psi(\boldsymbol{r}_1, \cdots, \boldsymbol{r}_J, \cdots, \boldsymbol{r}_I, \cdots, \boldsymbol{r}_N) = \psi(\boldsymbol{r}_1, \cdots, \boldsymbol{r}_I, \cdots, \boldsymbol{r}_J, \cdots, \boldsymbol{r}_N) \quad , \tag{5.3}$$

is equivalent to the system described by the free Hamiltonian

$$\tilde{H} = \sum_{I=1}^{N} \frac{1}{2m} \boldsymbol{p}_I^2 \quad , \tag{5.4}$$

acting on multi-valued wavefunctions $\tilde{\psi}$ such that

$$\tilde{\psi}(\boldsymbol{r}_1, \cdots, \boldsymbol{r}_J, \cdots, \boldsymbol{r}_I, \cdots, \boldsymbol{r}_N) = \mathrm{e}^{i\pi\nu} \, \tilde{\psi}(\boldsymbol{r}_1, \cdots, \boldsymbol{r}_I, \cdots, \boldsymbol{r}_J, \cdots, \boldsymbol{r}_N) \quad , \tag{5.5}$$

where

$$\nu = \frac{e^2}{2\pi \, \hbar \, c \, \kappa} \quad . \tag{5.6}$$

The two Hamiltonians (5.1) and (5.4) are simply related to each other by a gauge transformation which on the plane allows to remove completely the effective non-local vector potential $\boldsymbol{A}_I(\boldsymbol{r}_1, ... \boldsymbol{r}_N)$ [9]. The solution of either one of the two equivalent eigenvalue problems

$$H' \, \psi = E \, \psi \quad , \tag{5.7a}$$

$$\tilde{H} \, \tilde{\psi} = E \, \tilde{\psi} \quad , \tag{5.7b}$$

is highly non-trivial. In the first case the difficulty is due to the explicit presence of a long-range and non-local interaction, while in the second case the complexity arises from the complicated boundary conditions (5.5) that the multi-valued

[9] On a compact surface, like for example a sphere or a torus, one must carefully treat zero modes as well as possible topological components of the gauge fields (Iengo and Lechner 1990,1991,1992).

wavefunctions have to satisfy. A complete solution to (5.7) is known only for the two-body problem (Leinaas and Myrheim 1977; Wilczek 1982a,b; Jackiw 1990; Johnson and Canright 1990; Vercin 1991); since there are no self-interactions the one-body problem is trivial and only partial results are available at the moment for the N-body problem with $N \geq 3$. In the next chapter we will report on some recent progress in finding the solution to (5.7) for any N when external fields are present. Here instead, we point out that an equivalent approach to the N-body problem with identical particles is provided by second quantization where the Schroedinger eigenvalue equations (5.7) are replaced by a 2+1 (non-relativistic) quantum field theory.

To show this, let us consider a non-relativistic (complex) matter field $\phi(t, \boldsymbol{x})$ of mass m, which for definiteness we take to be bosonic [10], and let us minimally couple it to an abelian gauge field A_α with a Chern-Simons kinetic term (see Chapter 3). This system is described by the following action (Jackiw 1990; Jackiw and Pi 1990; Ezawa *et al.* 1991a,b; Ezawa and Iwazaki 1991)

$$\begin{aligned} S_0 &= \int dt\ \mathcal{L}_0 \\ &= \int d^3x \left[\mathrm{i}\phi^\dagger D_0\phi + \frac{1}{2m}\phi^\dagger \left({D_1}^2 + {D_2}^2\right)\phi + \frac{\kappa}{2}\epsilon^{\alpha\beta\gamma}A_\alpha\partial_\beta A_\gamma \right] \end{aligned} \tag{5.8}$$

where $D_\alpha = \partial_\alpha + \mathrm{i}\, e\, A_\alpha$ is the covariant derivative. (To avoid unnecessary redundancy in the formulas, from now on we set $c = \hbar = 1$, so that in particular $x^0 = t$. As far as indices are concerned, we keep instead the same conventions of Chapter 3.)

Varying the action S_0 with respect to A_α, we obtain

$$\epsilon^{\alpha\beta\gamma}F_{\beta\gamma} = \frac{2e}{\kappa}\, j^\alpha \tag{5.9}$$

where the current j^α is explicitly given by

$$\begin{aligned} j^0 &= \phi^\dagger\phi \equiv \rho \ , \\ j^i &= \frac{1}{2mi}\left(\phi^\dagger D_i\phi - (D_i\phi)^\dagger\, \phi\right) \ . \end{aligned} \tag{5.10}$$

ρ and $\boldsymbol{j}$ are respectively the number-density and the current-density operators and satisfy the continuity equation

$$\partial_t\rho + \nabla\cdot\boldsymbol{j} = 0 \ . \tag{5.11}$$

As is clear from (5.9), the Chern-Simons field strength $F_{\beta\gamma} = \partial_\beta A_\gamma - \partial_\gamma A_\beta$ is completely determined by the particle currents. Now we are going to demonstrate that also the Chern-Simons potential A_α itself is not an independent degree of freedom.

The $\alpha = 0$ component of (5.9) is

[10] Of course a similar and equivalent discussion can be done for fermionic matter fields.

$$B = -\frac{e}{\kappa}\,\rho \tag{5.12}$$

where $B = \nabla \wedge \boldsymbol{A} = -F_{12}$ is the Chern-Simons magnetic field. This equation clearly represents the second-quantized version of the quantum mechanical constraint (3.2.22) [11]. In the transverse gauge $\partial_i A^i = 0$, we can invert (5.12) without ambiguities and solve for the vector potential $\boldsymbol{A}$. Formally we get

$$A^i(x) = \epsilon^{ij}\frac{\partial}{\partial x^j}\left(\frac{e}{\kappa}\int d^2y\; G(\boldsymbol{x}-\boldsymbol{y})\,\rho(y)\right) \tag{5.13}$$

where G is the Green function for the Laplacian $\Delta = \nabla \cdot \nabla$, satisfying

$$\Delta G(\boldsymbol{x}-\boldsymbol{y}) = \delta(\boldsymbol{x}-\boldsymbol{y}) \quad . \tag{5.14}$$

As is well known, the explicit solution to (5.14) is

$$G(\boldsymbol{x}-\boldsymbol{y}) = \frac{1}{2\pi}\ln|\boldsymbol{x}-\boldsymbol{y}| \quad , \tag{5.15}$$

so that A^i can be written as follows

$$\begin{aligned} A^i(x) &= \epsilon^{ij}\frac{\partial}{\partial x^j}\left(\frac{e}{2\pi\kappa}\int d^2y\;\ln|\boldsymbol{x}-\boldsymbol{y}|\;\rho(y)\right) \\ &= -\frac{e}{2\pi\kappa}\int d^2y\;\frac{\partial}{\partial x^i}\varphi(\boldsymbol{x}-\boldsymbol{y})\,\rho(y) \quad , \end{aligned} \tag{5.16}$$

where

$$\varphi(\boldsymbol{x}-\boldsymbol{y}) = \tan^{-1}\left(\frac{x^2-y^2}{x^1-y^1}\right) \tag{5.17}$$

(see also (3.2.30) and (3.2.31)). Since the function $\varphi(\boldsymbol{x}-\boldsymbol{y})$ is multi-valued, particular care must be used in moving the derivative $\partial/\partial x^i$ out of the second integral in (5.16) and thus displaying $\boldsymbol{A}$ as a gradient. In fact, to eliminate the ambiguities deriving from the multi-valuedness of $\varphi(\boldsymbol{x}-\boldsymbol{y})$, one must fix a branch-cut in the $\boldsymbol{y}$-plane starting at $\boldsymbol{x}$. Whatever choice is made for this cut, the resulting range of integration of $\boldsymbol{y}$ will depend on $\boldsymbol{x}$, and hence extra contributions are produced in moving $\partial/\partial x^i$ outside the $\boldsymbol{y}$ integral. Therefore, it is not correct in general to write

$$\begin{aligned} \boldsymbol{A}(x) &= -\frac{e}{2\pi\kappa}\int d^2y\;\nabla\varphi(\boldsymbol{x}-\boldsymbol{y})\,\rho(y) \\ &= -\frac{e}{2\pi\kappa}\,\nabla\left(\int d^2y\;\varphi(\boldsymbol{x}-\boldsymbol{y})\,\rho(y)\right) \quad , \end{aligned} \tag{5.18}$$

and consequently it is not true in general that $\boldsymbol{A}$ is a pure gauge and can be removed by a gauge transformation. However there is a special situation where (5.18) is certainly correct and $\boldsymbol{A}$ is a pure gauge, namely the case in which the number-density operator $\rho(x)$ is a sum of δ-functions, as is appropriate for a collection

[11] Notice that here ρ is a matter density whereas in Chapter 3, ρ was a charge density; the latter is simply obtained from the matter density upon multiplication by e.

of non-relativistic, localized point-particles (Jackiw and Pi 1990). In the quantum field theory described by (5.8), we can always choose to remain consistently in such a case, provided that we do not consider the model as the non-relativistic limit of some relativistic quantum field theory. Indeed it is well-known that in a relativistic theory the eigenvalues of $\rho(x)$ are not sums of δ-functions because particles are not point-like but extended (see for instance (Forte 1992a,b)). In this case then, particular attention must be paid to the order in which the non-relativistic limit and the integrals (5.16) or (5.18) are computed (Dunne *et al.* 1992b).

The space components of (5.9) are

$$F_{i0} \equiv \partial_i A_0 - \partial_0 A_i = \frac{e}{\kappa}\epsilon_{ij} j^j \ , \tag{5.19}$$

which together with the continuity equation (5.11), allow to solve for the scalar potential A_0 yielding

$$A_0(x) = -\frac{e}{\kappa}\int d^2y \ G(\boldsymbol{x}-\boldsymbol{y}) \ \nabla \wedge \boldsymbol{j}(y) \ . \tag{5.20}$$

From (5.16) and (5.20), we explicitly see that the Chern-Simons gauge field A_α is entirely determined by the matter configuration, that is by ρ and $\boldsymbol{j}$.

The use of the Green function $G(\boldsymbol{x}-\boldsymbol{y})$ in expressions like (5.16) and (5.20) must be accompanied by a proper analysis of possible divergent and singular terms. For example, it is important to realize that

$$\epsilon^{ij}\frac{\partial}{\partial x^j}G(\boldsymbol{x}-\boldsymbol{y}) \tag{5.21}$$

is ill-defined at $\boldsymbol{x}=\boldsymbol{y}$. Therefore for a complete specification of the theory we must supplement (5.14) and (5.15) with a regularization prescription. Every such a regularization that preserves the antisymmetry under space reflections leads to a vanishing result for (5.21) at $\boldsymbol{x}=\boldsymbol{y}$. For definiteness it is convenient to use the following prescription

$$\epsilon^{ij}\frac{\partial}{\partial x^j}G(\boldsymbol{x}) \longrightarrow \epsilon^{ij}\frac{\partial}{\partial x^j}G^a(\boldsymbol{x}) \ , \tag{5.22}$$

where the regulated Green function $G^a(\boldsymbol{x})$ is

$$G^a(\boldsymbol{x}) = \frac{1}{a\,\pi}\int d^2z\left(\frac{1}{2\pi}\ln|\boldsymbol{x}-\boldsymbol{z}|\right)\mathrm{e}^{-\boldsymbol{z}^2/a} \ . \tag{5.23}$$

When the regulator parameter a is removed, we clearly get

$$\lim_{a\to 0} G^a(\boldsymbol{x}) = G(\boldsymbol{x}) = \frac{1}{2\pi}\ln|\boldsymbol{x}| \ . \tag{5.24}$$

On the other hand, it is easy to see that

$$\lim_{\boldsymbol{x}\to 0}\epsilon^{ij}\frac{\partial}{\partial x^j}G^a(\boldsymbol{x}) = 0 \tag{5.25}$$

for any a. If (5.22) is systematically used, all ambiguities are eliminated and the theory is completely specified.

Let us now quantize the action (5.8) by imposing equal-time commutation relations for the bosonic field ϕ, namely

$$\begin{aligned}[][\phi(\boldsymbol{x}), \phi^\dagger(\boldsymbol{y})] &= \delta(\boldsymbol{x}-\boldsymbol{y}) \ , \\ [\phi(\boldsymbol{x}), \phi(\boldsymbol{y})] &= [\phi^\dagger(\boldsymbol{x}), \phi^\dagger(\boldsymbol{y})] = 0 \ .\end{aligned} \tag{5.26}$$

(For simplicity we dropped the time arguments.) Since the gauge field $\boldsymbol{A}$ is a function of the number-density operator $\rho = \phi^\dagger\phi$, the commutation relation between $\boldsymbol{A}$ and ϕ is not trivial. In fact from (5.26) and (5.16), it is easy to derive that

$$[A^i(\boldsymbol{x}), \phi(\boldsymbol{y})] = -\frac{e}{\kappa}\epsilon^{ij}\frac{\partial}{\partial x^j}G(\boldsymbol{x}-\boldsymbol{y})\phi(\boldsymbol{y}) \ . \tag{5.27}$$

As pointed out earlier, one should really use (5.22) so that no ill-defined objects at $\boldsymbol{x}=\boldsymbol{y}$ appear. In particular, from (5.25) we conclude that

$$[A^i(\boldsymbol{x}), \phi(\boldsymbol{x})] = 0 \ . \tag{5.28}$$

Therefore there are no ordering ambiguities in the quantum Lagrangian (5.8) or in the Hamiltonian arising from it. The latter turns out to be

$$\begin{aligned} H_0 &= \int d^2x \ \mathcal{H}_0(\boldsymbol{x}) \ , \\ \mathcal{H}_0(\boldsymbol{x}) &= \frac{1}{2m}\Pi^\dagger(\boldsymbol{x}) \cdot \Pi(\boldsymbol{x}) \end{aligned} \tag{5.29}$$

where $\Pi(\boldsymbol{x})$ is the momentum operator associated to ϕ, namely

$$\Pi(\boldsymbol{x}) = [\nabla - \mathrm{i}e\boldsymbol{A}(\boldsymbol{x})]\phi(\boldsymbol{x}) \equiv D\phi(\boldsymbol{x}) \ , \tag{5.30}$$

and $\Pi^\dagger(\boldsymbol{x}) = (D\phi(\boldsymbol{x}))^\dagger$ is its hermitian conjugate. For later convenience we observe that, after an integration by parts, the Hamiltonian (5.29) can also be written in the following way

$$H_0 = \frac{1}{2m}\int d^2x \left[-\phi^\dagger\Delta\phi + \mathrm{i}e\boldsymbol{A}\cdot\left(\phi^\dagger\,\nabla\phi - \nabla\phi^\dagger\,\phi\right) + e^2\boldsymbol{A}^2\,\phi^\dagger\phi\right] \ . \tag{5.31}$$

Let us now derive the first-quantized Schroedinger equation (5.7a) from our second-quantized theory. First of all let us observe that the number operator

$$\hat{N} = \int d^2x \ \rho(\boldsymbol{x}) \ , \tag{5.32}$$

and the Hamiltonian H_0 in (5.29) commute. Therefore they can be simultaneously diagonalized as follows

$$\begin{aligned} H_0\,|E,N\rangle &= E\,|E,N\rangle \ , \\ \hat{N}\,|E,N\rangle &= N\,|E,N\rangle \ . \end{aligned} \tag{5.33}$$

Furthermore, we assume the existence of a vacuum state $|0\rangle$ annihilated by $\phi(\boldsymbol{x})$, H_0 and $\hat{N}$,

$$\begin{aligned} \phi(\boldsymbol{x})\,|0\rangle &= \langle 0|\,\phi^\dagger(\boldsymbol{x}) = 0 \;, \\ H_0\,|0\rangle &= \hat{N}\,|0\rangle = 0 \;, \end{aligned} \tag{5.34}$$

but such that $\phi^\dagger(\boldsymbol{x})$ acts non trivially on it. The states for a fixed number N of particles in configuration space are

$$|\boldsymbol{r}_1,\cdots,\boldsymbol{r}_N\rangle \equiv \phi^\dagger(\boldsymbol{r}_1)\cdots\phi^\dagger(\boldsymbol{r}_N)|0\rangle \;, \tag{5.35}$$

and consequently the N-body wavefunctions are

$$\psi(\boldsymbol{r}_1,\cdots,\boldsymbol{r}_N) = \langle \boldsymbol{r}_1,\cdots,\boldsymbol{r}_N|E,N\rangle = \langle 0|\phi(\boldsymbol{r}_1)\cdots\phi(\boldsymbol{r}_N)|E,N\rangle \;. \tag{5.36}$$

If we project the first equation in (5.33) onto the states (5.35), we get

$$\begin{aligned} \langle \boldsymbol{r}_1,\cdots,\boldsymbol{r}_N|\; H_0\;|E,N\rangle &= \langle 0|\phi(\boldsymbol{r}_1)\cdots\phi(\boldsymbol{r}_N)\; H_0\;|E,N\rangle \\ &= \langle 0|\;\left[\phi(\boldsymbol{r}_1)\cdots\phi(\boldsymbol{r}_N),H_0\right]\;|E,N\rangle \\ &= E\;\langle 0|\phi(\boldsymbol{r}_1)\cdots\phi(\boldsymbol{r}_N)|E,N\rangle \\ &= E\;\psi(\boldsymbol{r}_1,\cdots,\boldsymbol{r}_N) \;. \end{aligned} \tag{5.37}$$

The commutator in the second line of (5.37) can be computed without any problem. Here we present only the explicit calculation for $N=2$, and then infer from it the generalization to any N. First of all let us compute the commutator $[\phi(\boldsymbol{r}_I),H_0]$. Using (5.31) and the canonical commutation relations (5.26), one easily gets

$$\begin{aligned} [\phi(\boldsymbol{r}_I),H_0] =& -\frac{1}{2m}\Delta_I\phi(\boldsymbol{r}_I) + \frac{\mathrm{i}e}{2m}\left[\boldsymbol{A}(\boldsymbol{r}_I)\cdot\nabla_I\phi(\boldsymbol{r}_I) + \nabla_I\cdot\left(\boldsymbol{A}(\boldsymbol{r}_I)\phi(\boldsymbol{r}_I)\right)\right] \\ &+ \frac{e^2}{2m}\boldsymbol{A}(\boldsymbol{r}_I)^2\phi(\boldsymbol{r}_I) \\ &+ \frac{\mathrm{i}e^2}{2m\kappa}\int d^2x\;\epsilon^{ij}\frac{\partial}{\partial x^j}G(\boldsymbol{x}-\boldsymbol{r}_I)\phi(\boldsymbol{r}_I) \\ &\qquad\times\left(\phi^\dagger(\boldsymbol{x})\frac{\partial}{\partial x^i}\phi(\boldsymbol{x}) - \frac{\partial}{\partial x^i}\phi^\dagger(\boldsymbol{x})\,\phi(\boldsymbol{x})\right) \\ &+ \frac{e^3}{2m\kappa}\int d^2x\;\epsilon^{ij}\left[\frac{\partial}{\partial x^j}G(\boldsymbol{x}-\boldsymbol{r}_I)\phi(\boldsymbol{r}_I)A^i(\boldsymbol{x})\rho(\boldsymbol{x})\right. \\ &\qquad\left.+A^i(\boldsymbol{x})\frac{\partial}{\partial x^j}G(\boldsymbol{x}-\boldsymbol{r}_I)\phi(\boldsymbol{r}_I)\rho(\boldsymbol{x})\right] \;, \end{aligned} \tag{5.38}$$

where $\Delta_I = \nabla_I\cdot\nabla_I = (\partial/\partial x_I^i)(\partial/\partial x_I^i)$, and the κ-dependent terms arise from (5.27). The right hand side of (5.38) can be reorganized with the help of (5.20), yielding

$$\begin{aligned} [\phi(\boldsymbol{r}_I),H_0] =& -\frac{1}{2m}\Delta_I\phi(\boldsymbol{r}_I) + e\,A^0(\boldsymbol{r}_I)\,\phi(\boldsymbol{r}_I) \\ &+ \frac{e^4}{2m\kappa^2}\int d^2x\;\nabla G(\boldsymbol{x}-\boldsymbol{r}_I)\cdot\nabla G(\boldsymbol{x}-\boldsymbol{r}_I)\,\rho(\boldsymbol{x})\,\phi(\boldsymbol{r}_I) \;. \end{aligned} \tag{5.39}$$

The last term in this equation is a quantum correction arising from the reordering of the integrand in the last line of (5.38).

Since both A^0 and $\boldsymbol{A}$ involve expressions that contain the operator $\phi^\dagger$ standing on the left and therefore annihilating the vacuum $\langle 0|$, we immediately conclude that

$$\langle 0|[\phi(\boldsymbol{r}_I), H_0] = -\frac{1}{2m}\Delta_I \langle 0|\phi(\boldsymbol{r}_I) \ . \tag{5.40}$$

Let us now return to the calculation of (5.37) for $N = 2$. Using (5.40) for $I = 1$ and (5.39) for $I = 2$, after some simple algebra, we obtain

$$\langle 0|[\phi(\boldsymbol{r}_1)\phi(\boldsymbol{r}_2), H_0]|E, 2\rangle = -\frac{1}{2m}\sum_{I=1}^{2}(\nabla_I - \mathrm{i}e\boldsymbol{A}(\boldsymbol{r}_1, \boldsymbol{r}_2))^2 \langle 0|\phi(\boldsymbol{r}_1)\phi(\boldsymbol{r}_2)|E, 2\rangle \tag{5.41}$$

where

$$A_I^i(\boldsymbol{r}_1, \boldsymbol{r}_2) = \frac{e}{\kappa}\epsilon^{ij}\frac{\partial}{\partial x_I^j}\left(\sum_{J\neq I} G(\boldsymbol{r}_I - \boldsymbol{r}_J)\right) \ . \tag{5.42}$$

Using the explicit expression for the Green function given in (5.15), we can rewrite this equation and get

$$A_I^i(\boldsymbol{r}_1, \boldsymbol{r}_2) = \frac{e}{2\pi\kappa}\epsilon^{ij}\sum_{J\neq I}\frac{r_I^j - r_J^j}{|\boldsymbol{r}_I - \boldsymbol{r}_J|^2} \ . \tag{5.43}$$

Therefore for $N = 2$, (5.37) becomes

$$-\frac{1}{2m}\sum_{I=1}^{2}(\nabla_I - \mathrm{i}e\boldsymbol{A}_I(\boldsymbol{r}_1, \boldsymbol{r}_2))^2 \ \psi(\boldsymbol{r}_1, \boldsymbol{r}_2) = E\ \psi(\boldsymbol{r}_1, \boldsymbol{r}_2) \ . \tag{5.44}$$

This is the first-quantized Schroedinger equation for two particles in the non-local potential (5.43). The generalization to any N is now straightforward and the resulting system is described precisely by (5.1). In this way we have explicitly shown the equivalence of the field theory formulation and the first-quantized approach to the problem of N identical particles.

It is now interesting to rewrite the first-quantized Hamiltonian (5.1) in complex notation. As we explained in Chapter 4, for every particle we can introduce the complex coordinates

$$\begin{aligned} z_I &= x_I^1 + \mathrm{i}x_I^2 \ , \\ \bar{z}_I &= x_I^1 - \mathrm{i}x_I^2 \ , \end{aligned} \tag{5.45}$$

and the corresponding derivative operators

$$\begin{aligned} \partial_I &\equiv \frac{\partial}{\partial z_I} = \frac{1}{2}\left(\frac{\partial}{\partial x_I^1} - \mathrm{i}\frac{\partial}{\partial x_I^2}\right) \ , \\ \bar{\partial}_I &\equiv \frac{\partial}{\partial \bar{z}_I} = \frac{1}{2}\left(\frac{\partial}{\partial x_I^1} + \mathrm{i}\frac{\partial}{\partial x_I^2}\right) \ . \end{aligned} \tag{5.46}$$

Then we define

$$\begin{aligned} A_I &\equiv -\frac{1}{2}\left(A_I^1 - \mathrm{i}A_I^2\right) = -\frac{\mathrm{i}e}{4\pi\kappa}\sum_{J\neq I}\frac{1}{z_I - z_J} \ , \\ \bar{A}_I &\equiv -\frac{1}{2}\left(A_I^1 + \mathrm{i}A_I^2\right) = \frac{\mathrm{i}e}{4\pi\kappa}\sum_{J\neq I}\frac{1}{\bar{z}_I - \bar{z}_J} \ , \end{aligned} \tag{5.47}$$

where the second equalities follow from the explicit expression of $\boldsymbol{A}_I$, in such a way that the covariant derivative operators in complex notation turn out to be

$$\begin{aligned} D_I &\equiv \partial_I + \mathrm{i}eA_I = \partial_I + \frac{e^2}{4\pi\kappa}\sum_{J\neq I}\frac{1}{z_I - z_J} \ , \\ \bar{D}_I &\equiv \bar{\partial}_I + \mathrm{i}e\bar{A}_I = \bar{\partial}_I - \frac{e^2}{4\pi\kappa}\sum_{J\neq I}\frac{1}{\bar{z}_I - \bar{z}_J} \ . \end{aligned} \tag{5.48}$$

If we recall that

$$\bar{\partial}\ \frac{1}{z} = \partial\ \frac{1}{\bar{z}} = \pi\ \delta^{(2)}(z) \ , \tag{5.49}$$

it is straightforward to check that

$$[D_I, \bar{D}_I] = -\frac{e^2}{2\kappa}\sum_{J\neq I}\delta^{(2)}(z_I - z_J) \ . \tag{5.50}$$

Consequently the first-quantized Hamiltonian

$$H' = -\frac{1}{2m}\sum_{I=1}^{N}\left(\nabla_I - \mathrm{i}e\boldsymbol{A}_I(\boldsymbol{r}_1, \cdots, \boldsymbol{r}_N)\right)^2$$

can be rewritten as follows

$$\begin{aligned} H' &= -\frac{1}{m}\sum_{I=1}^{N}\left(D_I\,\bar{D}_I + \bar{D}_I\,D_I\right) \\ &= -\frac{2}{m}\sum_{I=1}^{N}\bar{D}_I\,D_I + \frac{e^2}{m\kappa}\sum_{I<J}\delta^{(2)}(z_I - z_J) \\ &= H_0' + \frac{e^2}{m\kappa}\sum_{I<J}\delta^{(2)}(z_I - z_J) \ . \end{aligned} \tag{5.51}$$

The total Hamiltonian H' differs from the normal ordered term

$$H_0' = -\frac{2}{m}\sum_{I=1}^{N}\bar{D}_I\,D_I = \frac{2}{m}\sum_{I=1}^{N}D_I^{\dagger}\,D_I \tag{5.52}$$

by a repulsive hard-core interaction among all pairs of particles. This δ-function interaction could be canceled by considering the action

$$S = S_0 + \frac{g}{2} \int d^3x \left(\phi^\dagger(x)\,\phi(x)\right)^2 \tag{5.53}$$

where S_0 is as in (5.8). The action S gives rise to the following quantum Hamiltonian

$$H = \int d^2x \left[\frac{1}{2m} \Pi^\dagger(x) \cdot \Pi(x) - \frac{g}{2} : \left(\phi^\dagger(x)\,\phi(x)\right)^2 : \right] \tag{5.54}$$

Using (5.10) and (5.12), the quartic interaction with coupling constant g that we have just introduced, may be written also as

$$\frac{g\kappa}{2e} : B(x)\rho(x) : \quad . \tag{5.55}$$

Displayed in this form, it is clear that it describes a magnetic-field – charge-density interaction. Going through the same procedure as we did before for $g = 0$, we can derive from (5.54) the Schroedinger equation and get the following first-quantized Hamiltonian

$$H' = \frac{2}{m} \sum_{I=1}^{N} D_I^\dagger D_I + \left(\frac{e^2}{m\kappa} - g \right) \sum_{I<J} \delta^{(2)}(z_I - z_J) \quad . \tag{5.56}$$

The quartic interaction of the field theory is equivalent to a δ-function potential in the quantum mechanical problem. If we fine-tune the coupling g and choose

$$g = \frac{e^2}{m\kappa} \quad , \tag{5.57}$$

then the hard-core interaction in (5.56) disappears and the Hamiltonian H' becomes simply the normal ordered one as in (5.52). It can be seen that the condition (5.57) on the coupling constant g plays a crucial role in studying non-trivial static solutions of the classical field theory described by (5.53). In particular it leads to the existence of self-dual solitons (Jackiw and Pi 1990; Dunne *et al.* 1991c; Dunne 1992). We mention also that the Hamiltonian (5.52) has been considered in (Girvin *et al.* 1990) as an example of an exactly solvable model for fractional statistics and was inspired by supersymmetry considerations (D_I and $D_I^\dagger$ resemble supersymmetry charges) as well as by the use of the Atiyah-Singer index theorem for particles in magnetic fields. Moreover, the relevance of (5.54) for the theory of the fractional quantum Hall Effect has been discussed in detail in (Ezawa *et al.* 1991a,b; Ezawa and Iwazaki 1991).

Let us now turn again to the second-quantized theory with $g = 0$. We have already remarked that in a non-relativistic context, the eigenvalues of the particle number-density operator ρ are sums of δ-functions, and therefore (5.17) is a correct formula. This implies that the space components of the gauge field A_α are a gradient

$$\begin{aligned} \boldsymbol{A}(x) &= \nabla \Lambda(x) \quad , \\ \Lambda(x) &= -\frac{e}{2\pi\kappa} \int d^2y \; \varphi(\boldsymbol{x} - \boldsymbol{y}) \; \rho(y) \quad . \end{aligned} \tag{5.58}$$

Now we want to show that also A_0 is a pure gauge, namely that

$$A_0(x) = -\partial_t \Lambda(x) \quad . \tag{5.59}$$

To show this, we first perform an integration by parts and rewrite (5.20) as follows

$$A_0(x) = \frac{e}{\kappa} \int d^2y \; \epsilon^{ij} \frac{\partial}{\partial x^j} G(\boldsymbol{x} - \boldsymbol{y}) \; j^i(y) \quad . \tag{5.60}$$

Then, upon using (5.14), we get

$$\begin{aligned} A_0(x) &= -\frac{e}{2\pi\kappa} \int d^2y \; \nabla\varphi(\boldsymbol{x} - \boldsymbol{y}) \cdot \boldsymbol{j}(y) \\ &= \frac{e}{2\pi\kappa} \int d^2y \; \nabla'\varphi(\boldsymbol{x} - \boldsymbol{y}) \cdot \boldsymbol{j}(y) \\ &= -\frac{e}{2\pi\kappa} \int d^2y \; \varphi(\boldsymbol{x} - \boldsymbol{y}) \; \nabla' \cdot \boldsymbol{j}(y) \\ &= \frac{e}{2\pi\kappa} \int d^2y \; \varphi(\boldsymbol{x} - \boldsymbol{y}) \; \partial_t \rho(y) = -\partial_t \Lambda(x) \quad . \end{aligned} \tag{5.61}$$

In deriving this result we have used the continuity equation (5.11) and have dropped surface terms in the integration by parts because, by assumption, the densities are localized.

Now we can combine (5.58) and (5.61) into a single covariant equation

$$A_\alpha(x) = -\partial_\alpha \Lambda(x) \quad , \tag{5.62}$$

which displays the Chern-Simons field as a pure gauge. Therefore it can be removed with a gauge transformation of parameter Λ, namely

$$A_\alpha \longrightarrow A'_\alpha = A_\alpha + \partial_\alpha \Lambda = 0 \quad . \tag{5.63}$$

Under such *singular* transformation, covariant derivatives turn into ordinary derivatives, and the action becomes simply

$$S'_0 = \int d^3x \left[\mathrm{i}\tilde{\phi}^\dagger \partial_0 \tilde{\phi} + \frac{1}{2m} \tilde{\phi}^\dagger \left({\partial_1}^2 + {\partial_2}^2 \right) \tilde{\phi} \right] \quad , \tag{5.64}$$

where the new matter fields $\tilde{\phi}$ are defined as

$$\begin{aligned} \tilde{\phi}(x) &= \mathrm{e}^{-\mathrm{i}e\Lambda(x)} \; \phi(x) \quad , \\ \tilde{\phi}^\dagger(x) &= \phi^\dagger(x) \; \mathrm{e}^{\mathrm{i}e\Lambda(x)} \quad . \end{aligned} \tag{5.65}$$

The action (5.64) is that of a free (complex) scalar field $\tilde{\phi}$; however such a field does *not* obey ordinary commutation relations. Indeed, using the equal-time commutator

$$[\phi(t, \boldsymbol{x}), \Lambda(t, \boldsymbol{y})] = -\frac{e}{2\pi\kappa} \varphi(\boldsymbol{y} - \boldsymbol{x}) \, \phi(t, \boldsymbol{x}) \quad , \tag{5.66}$$

which follows from (5.26) and (5.58), it is easy to show that for $\boldsymbol{x} \neq \boldsymbol{y}$

$$
\begin{aligned}
\tilde{\phi}(t,\boldsymbol{x})\tilde{\phi}(t,\boldsymbol{y}) &= \mathrm{e}^{-\mathrm{i}e\Lambda(t,\boldsymbol{x})}\phi(t,\boldsymbol{x})\mathrm{e}^{-\mathrm{i}e\Lambda(t,\boldsymbol{y})}\phi(t,\boldsymbol{y}) \\
&= \mathrm{e}^{-\mathrm{i}e[\Lambda(t,\boldsymbol{x})+\Lambda(t,\boldsymbol{y})]}\mathrm{e}^{\mathrm{i}\nu\varphi(\boldsymbol{y}-\boldsymbol{x})}\phi(t,\boldsymbol{x})\phi(t,\boldsymbol{y}) \\
&= \mathrm{e}^{-\mathrm{i}e[\Lambda(t,\boldsymbol{x})+\Lambda(t,\boldsymbol{y})]}\mathrm{e}^{\mathrm{i}\nu\varphi(\boldsymbol{y}-\boldsymbol{x})}\phi(t,\boldsymbol{y})\phi(t,\boldsymbol{x}) \\
&= \mathrm{e}^{-\mathrm{i}e[\Lambda(t,\boldsymbol{x})+\Lambda(t,\boldsymbol{y})]}\mathrm{e}^{\mathrm{i}\nu\varphi(\boldsymbol{y}-\boldsymbol{x})}\mathrm{e}^{\mathrm{i}e\Lambda(t,\boldsymbol{y})}\tilde{\phi}(t,\boldsymbol{y})\mathrm{e}^{\mathrm{i}e\Lambda(t,\boldsymbol{x})}\tilde{\phi}(t,\boldsymbol{x}) \\
&= \mathrm{e}^{\mathrm{i}\nu[\varphi(\boldsymbol{y}-\boldsymbol{x})-\varphi(\boldsymbol{x}-\boldsymbol{y})]}\tilde{\phi}(t,\boldsymbol{y})\tilde{\phi}(t,\boldsymbol{x}) \quad ,
\end{aligned}
\tag{5.67}
$$

where $\nu = e^2/(2\pi\kappa)$ as in (5.6). If $\boldsymbol{y}$ is moved counterclockwise around $\boldsymbol{x}$, then

$$
\varphi(\boldsymbol{y}-\boldsymbol{x}) - \varphi(\boldsymbol{x}-\boldsymbol{y}) = \pi \quad , \tag{5.68}
$$

as is clear from Fig. 5.1, and so we have

$$
\tilde{\phi}(t,\boldsymbol{x})\tilde{\phi}(t,\boldsymbol{y}) = \mathrm{e}^{\mathrm{i}\pi\nu}\tilde{\phi}(t,\boldsymbol{y})\tilde{\phi}(t,\boldsymbol{x}) \tag{5.69}
$$

Thus we have proved that the matter field $\tilde{\phi}$ obeys *anyonic* commutation relations of statistics ν. A calculation similar to (5.67) leads to

$$
\tilde{\phi}(t,\boldsymbol{x})\tilde{\phi}^\dagger(t,\boldsymbol{y}) = \mathrm{e}^{\mathrm{i}\nu[\varphi(\boldsymbol{x}-\boldsymbol{y})-\varphi(\boldsymbol{y}-\boldsymbol{x})]}\ \tilde{\phi}^\dagger(t,\boldsymbol{y})\tilde{\phi}(t,\boldsymbol{x}) + \delta(\boldsymbol{x}-\boldsymbol{y}) \quad . \tag{5.70}
$$

If $\boldsymbol{x} \neq \boldsymbol{y}$, we obtain

$$
\tilde{\phi}(t,\boldsymbol{x})\tilde{\phi}^\dagger(t,\boldsymbol{y}) = \mathrm{e}^{-\mathrm{i}\pi\nu}\tilde{\phi}^\dagger(t,\boldsymbol{y})\tilde{\phi}(t,\boldsymbol{x}) \quad , \tag{5.71}
$$

which exhibits again anyonic statistics ν; however we remark that for $\boldsymbol{x} = \boldsymbol{y}$ the canonical commutator remains unchanged since the phase proportional to ν in this case is vanishing.

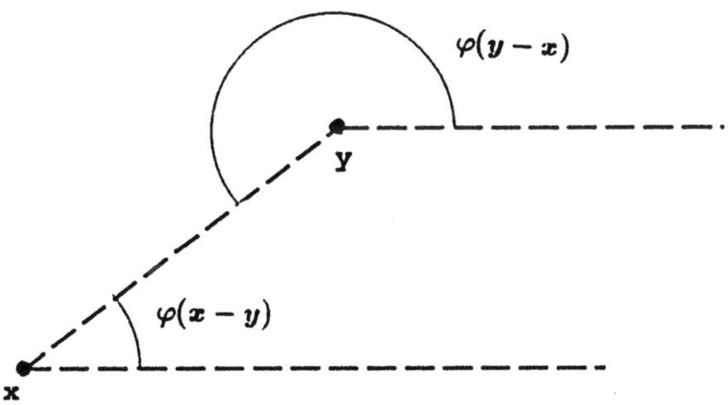

Fig. 5.1. $\varphi(\boldsymbol{y}-\boldsymbol{x}) - \varphi(\boldsymbol{x}-\boldsymbol{y}) = \pi$.

The non-relativistic field theory described by the action S_0 in (5.8) or by the action S in (5.53) is invariant under the usual space-time symmetries: time translations, space translations, space rotations and Galilean boosts. To each of these symmetries one can associate a conserved Noether current and a corresponding conserved Noether charge according to the standard procedure. We refer the reader to (Jackiw and Pi 1990) for a complete treatment of such symmetries and the explicit derivation of the Noether currents. Here we simply focus on the rotational invariance and the associated angular momentum, which turn out to be (see also (3.2.51))

$$J = m \int d^2x \; \epsilon^{ij} x^i \, j^j \tag{5.72}$$

where the current $\boldsymbol{j}$ is defined in (5.10). Using the explicit expression of $\boldsymbol{j}$, we obtain

$$\begin{aligned} J &= \frac{1}{2\mathrm{i}} \int d^2x \; \epsilon^{ij} x^i \left[\phi^\dagger(x) \left(\partial_j + \mathrm{i} e A_j(x) \right) \phi(x) - \left(\partial_j - \mathrm{i} e A_j(x) \right) \phi^\dagger(x) \, \phi(x) \right] \\ &= \frac{1}{2\mathrm{i}} \int d^2x \; \epsilon^{ij} x^i \left[\phi^\dagger(x) \partial_j \phi(x) - \partial_j \phi^\dagger(x) \, \phi(x) \right] + e \int d^2x \; \epsilon^{ij} x^i A_j(x) \rho(x) \end{aligned} \tag{5.73}$$

where $\rho = \phi^\dagger \phi$.

Upon setting $\phi = \mathrm{e}^{\mathrm{i}\theta} \rho^{1/2}$ and $\phi^\dagger = \mathrm{e}^{-\mathrm{i}\theta} \rho^{1/2}$, we see that

$$\frac{1}{2\mathrm{i}} \left[\phi^\dagger(x) \partial_j \phi(x) - \partial_j \phi^\dagger(x) \, \phi(x) \right] = \partial_j \theta \, \rho \;\; . \tag{5.74}$$

Therefore, if we assume that the matter fields have a configuration which vanishes rapidly enough at spatial infinity [12], then the first integral in (5.73) vanishes and the angular momentum is simply

$$J = e \int d^2x \; \epsilon^{ij} x^i A_j(x) \rho(x) \;\; . \tag{5.75}$$

Now we have to insert in this formula the expression for A_j given in (5.13) which we rewrite here for convenience

$$\begin{aligned} A_j(x) = -A^j(x) &= -\frac{e}{\kappa} \int d^2y \; \epsilon^{jk} \frac{\partial}{\partial x^k} G(\boldsymbol{x} - \boldsymbol{y}) \, \rho(y) \\ &= -\frac{2\pi\nu}{e} \int d^2y \; \epsilon^{jk} \frac{\partial}{\partial x^k} G^a(\boldsymbol{x} - \boldsymbol{y}) \, \rho(y) \bigg|_{a \to 0} \;\; . \end{aligned} \tag{5.76}$$

The last equality follows from (5.6) and (5.22). Combining everything together we obtain

$$\begin{aligned} J &= -2\pi\nu \int d^2x \int d^2y \; \rho(x) \epsilon^{ij} x^i \epsilon^{jk} \frac{\partial}{\partial x^k} G^a(\boldsymbol{x} - \boldsymbol{y}) \rho(y) \bigg|_{a \to 0} \\ &= 2\pi\nu \int d^2x \int d^2y \; \rho(x) x^i \frac{\partial}{\partial x^i} G^a(\boldsymbol{x} - \boldsymbol{y}) \rho(y) \bigg|_{a \to 0} \;\; . \end{aligned} \tag{5.77}$$

Let us evaluate this expression for the matter density given by

[12] This is certainly the case for non-relativistic configurations, where ρ is a sum of δ-functions.

$$\rho(x) = \sum_{I=1}^{N} \delta^{(2)}(\boldsymbol{x} - \boldsymbol{r}_I(t)) \ . \tag{5.78}$$

As we have already remarked, this is equivalent to consider N point-like non-relativistic particles located at $\boldsymbol{r}_I(t)$. Inserting (5.78) into (5.77), we have

$$\begin{aligned} J &= 2\pi\nu \sum_{I,J=1}^{N} \int d^2x \int d^2y \ \delta^{(2)}(\boldsymbol{x} - \boldsymbol{r}_I(t)) x^i \frac{\partial}{\partial x^i} G^a(\boldsymbol{x} - \boldsymbol{y}) \, \delta^{(2)}(\boldsymbol{y} - \boldsymbol{r}_J(t)) \Big|_{a \to 0} \\ &= 2\pi\nu \sum_{I,J=1}^{N} r_I^i \frac{\partial}{\partial r_I^i} G^a(\boldsymbol{r}_I - \boldsymbol{r}_J) \Big|_{a \to 0} \\ &= 2\pi\nu \sum_{I=1}^{N} \sum_{J \neq I} r_I^i \frac{\partial}{\partial r_I^i} G^a(\boldsymbol{r}_I - \boldsymbol{r}_J) \Big|_{a \to 0} \end{aligned} \tag{5.79}$$

where the last step is due to (5.25), that is to the absence of self-interactions in the regulated theory. We can proceed even further if we observe that the regulator a can now be removed from (5.79), so that using the explicit expression for G given in (5.15), we obtain

$$\begin{aligned} J &= 2\pi\nu \sum_{I=1}^{N} \sum_{J \neq I} r_I^i \frac{\partial}{\partial r_I^i} G(\boldsymbol{r}_I - \boldsymbol{r}_J) \Big|_{a \to 0} \\ &= \nu \sum_{I=1}^{N} \sum_{J \neq I} r_I^i \frac{r_I^i - r_J^i}{|\boldsymbol{r}_I - \boldsymbol{r}_J|^2} = \nu \frac{1}{2} N(N-1) \ . \end{aligned} \tag{5.80}$$

This result coincides with the one we have derived in Chapter 3 using the first-quantized theory (cf. (3.2.53)). Notice that for point-like particles there are no self interactions and the angular momentum J comprises only the *orbital* part, as is clear from the fact that (5.80) vanishes for $N = 1$. We can therefore conclude that in a non-relativistic theory with Chern-Simons gauge fields, there is a fractional orbital angular momentum and particles carry no spin.

The situation is drastically different in a relativistic field theory (Forte and Jolicoeur 1991; Forte 1992a,b) or even when the model emerges from a relativistic theory in the non-relativistic limit. In this latter case, one should consider extended particles in the limit as their size approaches zero; then a delicate question arises about the order in which the size of the particles and the regulator parameter a are taken to zero. If we perform the zero-size limit first, and then compute J and remove the regulator at the end, we clearly obtain the result (5.80); however if we proceed in the reverse order and first compute J for extended densities and then take the limit of zero-size, we get a different result (Dunne *et al.* 1992b) . Now we show explicitly how this may happen.

Let us consider a matter density of the form

$$\rho(x) = \sum_{I=1}^{N} \rho_I(x) \tag{5.81}$$

where $\rho_I(x)$ is a spherically symmetric function centered around the point $\boldsymbol{r}_I(t)$, which in the limit of point-like particles eventually reduces to a δ-function. For example we could take

$$\rho_I(x) = \frac{1}{b\pi} \mathrm{e}^{-(\boldsymbol{x}-\boldsymbol{r}_I(t))^2/b} \ , \tag{5.82}$$

such that

$$\begin{aligned} \int d^2x \ \rho_I(x) &= 1 \ , \\ \lim_{b\to 0} \rho_I(x) &= \delta^{(2)}(\boldsymbol{x} - \boldsymbol{r}_I(t)) \ . \end{aligned} \tag{5.83}$$

The parameter b can be considered as related to the size of the particle, and will be set to zero at the very end of the calculation. Upon inserting (5.81) into (5.77), we get

$$\begin{aligned} J &= 2\pi\nu \sum_{I,J=1}^{N} \int d^2x \int d^2y \ \rho_I(x) x^i \frac{\partial}{\partial x^i} G^a(\boldsymbol{x} - \boldsymbol{y}) \rho_J(y) \Big|_{a\to 0}\Big|_{b\to 0} \\ &\equiv L_\nu + S_\nu \ , \end{aligned} \tag{5.84}$$

where

$$\begin{aligned} L_\nu &= 2\pi\nu \sum_{I=1}^{N} \sum_{J\neq I} \int d^2x \int d^2y \ \rho_I(x) x^i \frac{\partial}{\partial x^i} G^a(\boldsymbol{x} - \boldsymbol{y}) \rho_J(y) \Big|_{a\to 0}\Big|_{b\to 0} \ , \\ S_\nu &= 2\pi\nu \sum_{I=1}^{N} \int d^2x \int d^2y \ \rho_I(x) x^i \frac{\partial}{\partial x^i} G^a(\boldsymbol{x} - \boldsymbol{y}) \rho_I(y) \Big|_{a\to 0}\Big|_{b\to 0} \ . \end{aligned} \tag{5.85}$$

The reason for this decomposition and the choice of the symbols will become apparent in a moment. The calculation of L_ν in (5.85) presents no difficulty, yielding as before

$$L_\nu = \nu \frac{1}{2} N(N-1) \ . \tag{5.86}$$

Comparing this with (5.80), we can conclude that in this calculation the order in which the limits $a \to 0$ and $b \to 0$ are taken, does not influence the result. However the situation is different for S_ν in (5.85). To see this, first of all let us observe that using the obvious property

$$\frac{\partial}{\partial x^i} G^a(\boldsymbol{x} - \boldsymbol{y}) = -\frac{\partial}{\partial y^i} G^a(\boldsymbol{x} - \boldsymbol{y}) \ , \tag{5.87}$$

S_ν can be rewritten as follows

$$S_\nu = \pi\nu \sum_{I=1}^{N} \int d^2x \int d^2y \; \rho_I(x)(x^i - y^i)\frac{\partial}{\partial x^i} G^a(x-y)\rho_I(y)\Big|_{a\to 0}\Big|_{b\to 0} \quad . \tag{5.88}$$

Then, using (5.82) and shifting the integration variables, we obtain

$$\begin{aligned} S_\nu &= \pi\nu \sum_{I=1}^{N} \int d^2x \int d^2y \; \frac{1}{b\pi} \mathrm{e}^{-x^2/b}(x^i - y^i)\frac{\partial}{\partial x^i} G^a(x-y) \; \frac{1}{b\pi}\mathrm{e}^{-y^2/b}\Big|_{a\to 0}\Big|_{b\to 0} \\ &= \frac{1}{2}\nu \sum_{I=1}^{N} \int d^2x \int d^2y \; \frac{1}{b\pi}\mathrm{e}^{-x^2/b} \; \frac{1}{b\pi}\mathrm{e}^{-y^2/b}\Big|_{b\to 0} = \frac{1}{2}\nu \; N \; . \end{aligned} \tag{5.89}$$

We stress once again that if we had computed (5.88) by taking first the limit $b \to 0$ and then the limit $a \to 0$, we would have gotten a zero result.

It is now clear what is the meaning of L_ν in (5.86) and of S_ν in (5.89). L_ν is a fractional *orbital* angular momentum, whereas S_ν is a fractional *spin*. L_ν, which is the second quantized version of the quantum mechanical angular momentum ℓ_ν (see (3.2.53)), originates from the interaction among different particles and is proportional to the number of pairs, $N(N-1)/2$. On the contrary S_ν, which does not vanish for $N = 1$, comes from self-interactions of *extended* anyons and is proportional to the number of particles, N. The combination

$$J = L_\nu + S_\nu = \frac{1}{2}\nu \; N(N-1) + \frac{1}{2}\nu \; N \tag{5.90a}$$

is often presented in the literature as

$$J = \frac{1}{2}\nu \; N^2 \quad , \tag{5.90b}$$

and is regarded as a manifestation of a non-standard composition rule for the angular momentum of particles with fractional statistics. Indeed, if one anyon has angular momentum $\nu/2$ – in agreement with the spin-statistics connection – N anyons have a total angular momentum $N^2\nu/2$. However we should realize that in $N^2\nu/2$ there is an orbital contribution and a true spin contribution, and if we keep this in mind, (5.90b) does not come as a surprise at all.

We can summarize our results as follows. For N point-like particles of statistics ν (see Fig. 5.2), the total angular momentum is

$$J = L_\nu = \frac{1}{2}\nu \; N(N-1) \quad , \tag{5.91}$$

and no spin is present.

For N extended anyons in the point-like limit (see Fig. 5.3), the total angular momentum is

$$J = L_\nu + S_\nu = \frac{1}{2}\nu \; N(N-1) + \frac{1}{2}\nu \; N \tag{5.92}$$

where S_ν is the spin and the standard spin-statistics relation is verified.

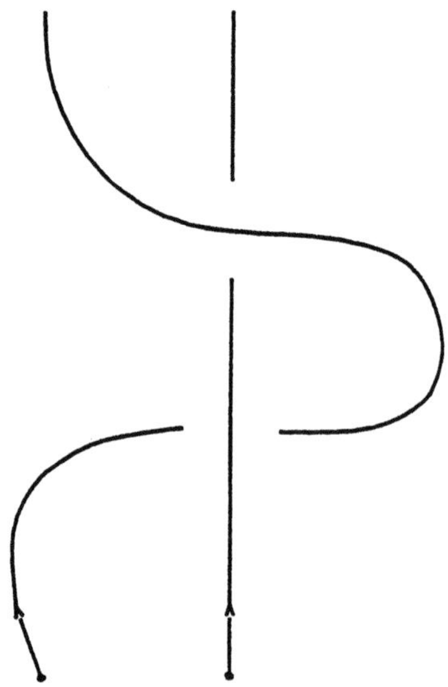

Fig. 5.2. Point-like anyons can be represented as particles with infinitesimally thin flux tubes. The fractional orbital angular momentum L_ν originates from the interaction between the flux tubes of different particles.

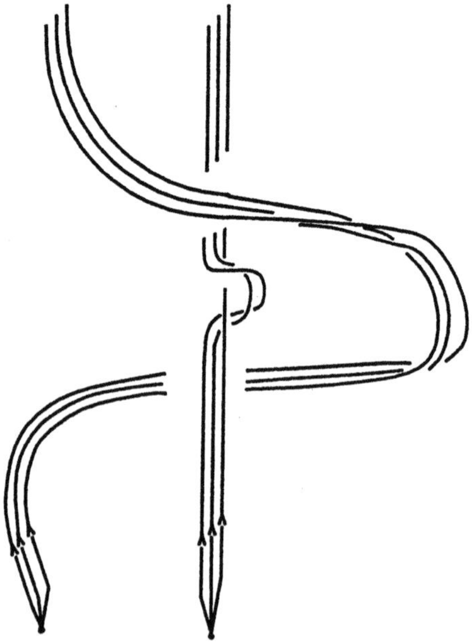

Fig. 5.3. Extended anyons can be represented as particles with extended flux tubes. In this case self-intersections occur and give rise to the spin $S\nu$.

6. Anyons in a Magnetic Field

The general theory of fractional statistics we have presented in the previous chapters, shows that anyons provide a fascinating model in which to study topological effects in quantum mechanics and in quantum field theory. To complete the analysis and see whether or not anyons are relevant for physics, one needs to develop the many-body theory, and then study the thermodynamic properties of systems of *very* large numbers of anyons as we customarily do for many-boson and many-fermion systems. However a deep understanding of the many-body problem for anyons is still lacking, despite the significant progress that has been recently made in various models using analytical, approximate and numerical methods.

It has been known since the early days of fractional statistics that the two-body anyon problem is completely soluble (Leinaas and Myrheim 1977; Wilczek 1982a,b; Wu 1984a) and that in the three-anyon problem with a harmonic potential a set of exact wavefunctions can be easily found by solving a partial differential equation (Wu 1984b). The latter approach has recently been extended to the N-anyon problem in the presence of a harmonic potential (Chou 1991a). On the other hand, the (related) problem of anyons in an external magnetic field has been analyzed by M.D. Johnson and G.S. Canright, who gave detailed results for the two-body problem as well as very partial results for the N-body problem (Johnson and Canright 1990). The external magnetic field breaks the free particle degeneracy and reveals a spectral structure reminiscent of the familiar Landau levels of charged particles in an external magnetic field (see for example (Landau and Lifschitz 1977)). This N-body problem has recently been studied in more detail (Dunne *et al.* 1991a,b) in connection with the surprising relationship between anyonic theories and conformal field theories which has been pointed out by many authors (Fubini 1991; Stone 1991a; Fubini and Lütken 1991; Moore and Read 1991; Cristofano *et al.* 1991a,b,c). Finally, we mention that numerical studies have recently been performed in the three- and four-anyon problems with a harmonic potential (Sporre *et al.* 1991a; Murphy *et al.* 1991; Sporre *et al.* 1991b), and that perturbative calculations for statistics infinitesimally close to the fermionic or bosonic ones are now available for many models (Chou 1991b; Mc Cabe and Ouvry 1991; Comtet *et al.* 1991; Karlhede and Westerberg 1991; Khare and Mc Cabe 1991; Dasnières de Veigy and Ouvry 1991; Sporre *et al.* 1991c; Chou *et al.* 1992).

In this chapter, I am going to present a systematic discussion of the N-body problem for anyons in the presence of an external magnetic field, following closely the article I wrote in collaboration with G. Dunne, S. Sciuto and C. Trugenberger (Dunne *et al.* 1992a). The study of anyons in magnetic fields is motivated by the following considerations. The only known example of objects which can be described as anyons are the quasi-particles and the quasi-holes of the fractional quantum Hall effect (see for instance (Prange and Girvin 1990) and Chapter 8). These have the important feature of feeling the underlying electrons exactly as a charged particle would feel an external uniform magnetic field. Thus in their physical realizations, anyons experience effective magnetic fields (not to mention the strong external magnetic field which drives the Hall effect itself). To find the solution to the problem of anyons in magnetic fields is therefore interesting and important also for practical purposes (for related works see also (Grundberg *et al.* 1991b; Karlhede and Westerberg 1991; Cho and Rim 1992).

6.1 The Eigenvalue Problem

The Hamiltonian for a collection of N identical anyons (labeled by capital Latin indices $I, J, \ldots$) of statistics ν, mass m and electric charge e, moving on a plane in the presence of an external constant, uniform magnetic field B (perpendicular to the plane) is given by [13]

$$H' = -\frac{1}{2m}\sum_{I=1}^{N}\left(\nabla_I - \mathrm{i}\boldsymbol{A}_I(\boldsymbol{r}_1,...\boldsymbol{r}_N) - \mathrm{i}e\boldsymbol{A}\left(\boldsymbol{r}_I\right)\right)^2 \quad . \tag{6.1.1}$$

We choose the external vector potential $\boldsymbol{A}(\boldsymbol{r}_I)$ in the symmetric gauge, namely

$$A^i\left(\boldsymbol{r}\right) = -\frac{B}{2}\epsilon^{ij}x^j \quad , \tag{6.1.2}$$

and without any loss of generality we assume B to be positive. The non-local effective gauge potential

$$A_I^i\left(\boldsymbol{r}_1,\ldots,\boldsymbol{r}_N\right) = \nu\sum_{J\neq I}\epsilon^{ij}\frac{r_I^j - r_J^j}{\left|\boldsymbol{r}_I - \boldsymbol{r}_J\right|^2} \tag{6.1.3}$$

describes the statistical interactions (see (3.2.27)). The associated statistical magnetic field B_I is

$$B_I\left(\boldsymbol{r}_1,\ldots,\boldsymbol{r}_N\right) = \nabla_I \wedge \boldsymbol{A}_I(\boldsymbol{r}_1,...\boldsymbol{r}_N) = -2\pi\,\nu\sum_{J\neq I}\delta^2\left(\boldsymbol{r}_I - \boldsymbol{r}_J\right) \quad , \tag{6.1.4}$$

so that each particle sees the $(N-1)$ others as vortices of strength $(-2\pi\,\nu)$, and a statistical phase $\pi\,\nu$ under interchange of two particles is acquired via the Aharonov–Bohm effect. Note that since the statistical phase is $\pi\,\nu$ we may restrict ν to values in the interval $[0,2)$.

The non-local potential $\boldsymbol{A}_I$ is pure gauge except at the locations of the particles,

$$\begin{aligned} A_I^i\left(\boldsymbol{r}_1,\ldots,\boldsymbol{r}_N\right) &= -\nu\frac{\partial}{\partial r_I^i}\sum_{J\neq I}\varphi_{IJ} \quad , \\ \varphi_{IJ} &= \tan^{-1}\left(\frac{x_I^2 - x_J^2}{x_I^1 - x_J^1}\right) \quad , \end{aligned} \tag{6.1.5}$$

[13] In this chapter we set $\hbar = c = 1$.

and therefore can be removed from (6.1.1) by going to the anyon gauge (see Chapter 4). The resulting Hamiltonian is

$$\tilde{H} = -\frac{1}{2m}\sum_{I=1}^{N}\left(\nabla_I - \mathrm{i}e\boldsymbol{A}\left(\boldsymbol{r}_I\right)\right)^2 \tag{6.1.6}$$

which acts on the redefined wavefunctions

$$\tilde{\psi}\left(\boldsymbol{r}_1,\ldots,\boldsymbol{r}_N\right) = \mathrm{e}^{\mathrm{i}\nu\sum_{I<J}\varphi_{IJ}}\;\psi\left(\boldsymbol{r}_1,\ldots,\boldsymbol{r}_N\right) \quad . \tag{6.1.7}$$

Thus in the anyon gauge, the Hamiltonian reduces to the usual sum of one-body Hamiltonians, the statistical interactions disappear and become entirely hidden in the boundary conditions which make the wavefunctions $\tilde{\psi}$ *multi-valued*. Consequently, the many-body Hilbert space is not a simple direct product of one-body Hilbert spaces. Throughout the rest of this chapter we shall work with these multi-valued wavefunctions, and for simplicity we drop the ˜ sign from $\tilde{\psi}$ and $\tilde{H}$.

Using the complex notation

$$\begin{aligned} z &= x^1 + \mathrm{i}\,x^2 \ , & \bar{z} &= x^1 - \mathrm{i}\,x^2 \ , \\ \partial &= \frac{\partial}{\partial z} \ , & \bar{\partial} &= \frac{\partial}{\partial \bar{z}} \ , \end{aligned} \tag{6.1.8}$$

the Hamiltonian becomes

$$H = \sum_{I=1}^{N}\left(-\frac{2}{m}\bar{\partial}_I\partial_I + \frac{e^2B^2}{8m}|z_I|^2\right) - \frac{eB}{2m}J_{\mathrm{c}} \ , \tag{6.1.9}$$

where J_{c} is the total (canonical) angular momentum operator

$$J_{\mathrm{c}} = \sum_{I=1}^{N}\left(z_I\partial_I - \bar{z}_I\bar{\partial}_I\right) \quad . \tag{6.1.10}$$

We note that J_{c} commutes with H so that we can solve the following eigenvalue equations

$$\begin{aligned} H\,\psi &= E\,\psi \ , \\ J_{\mathrm{c}}\,\psi &= J\,\psi \ . \end{aligned} \tag{6.1.11}$$

For $N = 1$ these equations form the Landau problem whose solution is known since the early days of quantum mechanics (Landau 1930). Let us now briefly review some elementary facts concerning this solution.

The properly normalized one-body Landau wavefunctions are

$$\psi_n^j(z,\bar{z}) = \sqrt{\frac{n!}{\pi\,2^{j+1}\,(n+j)!}}\; z^j\,L_n^j\left(\frac{eB}{2}|z|^2\right)\exp\left(-\frac{eB}{4}|z|^2\right) \tag{6.1.12}$$

where n and j are integers such that $n \geq 0$ and $j \geq -n$. They satisfy (6.1.11) with

$$E = \omega \left(n + \frac{1}{2} \right) \ ,$$
$$J = j \ . \tag{6.1.13}$$

where $\omega \equiv eB/m$ is the cyclotron frequency and the L^j_n are generalized Laguerre polynomials. We refer the reader to Appendix A where we list some properties of the Laguerre polynomials that are useful for our discussion. Here we remark in particular that when j is negative, which can happen only for excited states $(n \geq 1)$,

$$z^j L^j_n \left(\frac{eB}{2} |z|^2 \right) \propto \bar{z}^{-j} L^{-j}_{n+j} \left(\frac{eB}{2} |z|^2 \right) \ . \tag{6.1.14}$$

Thus an antianalytic prefactor can appear only for excited states. The energy spectrum is structured in discrete levels similar to those of the harmonic oscillator and known as Landau levels (see for example (Landau and Lifschitz 1977)). These levels have an infinite degeneracy given by all states with the same quantum number n and arbitrary angular momentum $j \geq -n$. However for samples of *finite* area this degeneracy is finite since in this case the angular momentum is bounded from above.

Given the wavefunctions (6.1.12), the solution of the N-body problem is immediate for bosons or fermions: in fact, all many-body wavefunctions are simply obtained by forming respectively all totally symmetric or antisymmetric products of N one-body wavefunctions. In particular, it is easy to show that the ground state for N particles of ordinary statistics has energy E and angular momentum J given by

$$\text{bosons:} \qquad E = \omega \frac{N}{2} \ , \qquad J = d_+ \ ,$$
$$\text{fermions:} \qquad E = \omega \frac{N}{2} \ , \qquad J = \frac{1}{2} N(N-1) + d_+ \ , \tag{6.1.15}$$

where d_+ is a non-negative integer. These are of course the expected results.

Let us now turn to the N-body problem for particles of fractional statistics. It is convenient to extract from the wavefunction in (6.1.11) the factor $\exp\left(-\frac{eB}{4} \sum_{I=1}^{N} |z_I|^2 \right)$ so that the eigenvalue problem reduces to

$$\hat{H} \hat{\psi} = \left(E - \omega \frac{N}{2} \right) \hat{\psi} \ , \tag{6.1.16a}$$

$$J_c \hat{\psi} = J \hat{\psi} \ , \tag{6.1.16b}$$

where the new Hamiltonian $\hat{H}$ and the new wavefunction $\hat{\psi}$ are defined by

$$\hat{H} = \sum_{I=1}^{N} \left[-\frac{2}{m} \bar{\partial}_I \partial_I + \frac{eB}{m} \bar{z}_I \bar{\partial}_I \right] \ ,$$
$$\hat{\psi} = \exp\left(\frac{eB}{4} \sum_{I}^{N} |z_I|^2 \right) \psi \ . \tag{6.1.17}$$

The wavefunction $\hat{\psi}$ must satisfy some physical requirements. First of all, $\hat{\psi}$ must vanish at points of coincidence if $\nu \neq 0$ (cf also (Grundberg *et al.* 1991a)). This hard-core requirement excludes the possibility of overlapping particles: indeed as we have seen in Chapter 2, the possibility of intermediate statistics in two space dimensions is due to the non-trivial topology of the configuration space of N identical particles with the *coincidence points excluded* and with permuted configurations identified, whose fundamental group is the braid group B_N (see (2.11) and (2.12)). According to our construction the wavefunction $\hat{\psi}$ must satisfy the multi-valued boundary conditions (6.1.7) which indeed characterize an Abelian representation of B_N.

A very general form of $\hat{\psi}$ with the above properties is

$$\hat{\psi} = \prod_{I<J} (z_{IJ})^{a_{IJ}} (\bar{z}_{IJ})^{b_{IJ}} \phi(z_1, \ldots, z_N; \bar{z}_1, \ldots, \bar{z}_N) + \text{symm} \tag{6.1.18}$$

where $z_{IJ} \equiv (z_I - z_J)$, "symm" means symmetrization with respect to permutations of the index I in the variables $z_I, \bar{z}_I$, and the parameters a_{IJ} and b_{IJ} satisfy

$$a_{IJ} - b_{IJ} = \nu \; [\text{mod}\, 2] \quad , \qquad a_{IJ} + b_{IJ} \geq 0 \quad , \tag{6.1.19}$$

for any pair (IJ). Finally, ϕ is a single-valued function which we *choose* to be regular and non-vanishing at all coincidence points.

A few remarks are in order. First of all, we observe that contrary to naive expectations, symmetrization does *not* automatically produce a bosonic wavefunction for $\nu \neq 0, 1$ as one can see from the simplest example

$$(z_{12})^\nu + \text{symm} = (z_{12})^\nu + (z_{21})^\nu = \left(1 + e^{i\pi\nu}\right)(z_{12})^\nu \quad . \tag{6.1.20}$$

Secondly, we want to emphasize that the restriction on the function ϕ in (6.1.18) to be regular and non-vanishing, amounts to the requirement that ϕ be expandable as Taylor series in z_{IJ} and $\bar{z}_{IJ}$ for all pairs (IJ), and that the leading term in this expansion be non vanishing. Moreover, since ϕ is by assumption single-valued, it is clear what is the meaning of the two conditions (6.1.19): the first one enforces the right statistics on $\hat{\psi}$; the second excludes possible divergences (notice however that a_{IJ} and b_{IJ} are not necessarily positive).

We want to stress that (6.1.18) is *not* the most general form of a wavefunction that is not singular at coincident points and has the right symmetry properties under exchanges. In fact, all functions with unusual angular dependence are excluded from our analysis. For example for $N = 3$ the function

$$(z_{12}\, z_{23}\, z_{13})^\nu \sqrt{|z_{12}|^2 + |z_{23}|^2 + |z_{13}|^2} \tag{6.1.21}$$

has the right braid group properties, is not singular for $z_{IJ} \to 0$ but is *not* of the form (6.1.18) because the square root term is not expandable as Taylor series in z_{IJ}! Clearly (6.1.21) is not a solution to our eigenvalue problem (6.1.16a) but is an example of those acceptable wavefunctions which are left out from our discussion and which could give rise to missing states. At the present time it is *not* known how to solve the eigenvalue problem for $N \geq 3$ anyons in its full generality.

Indeed, the exact analysis we are going to present in the following, based only on wavefunctions of the form (6.1.18), is the most extensive one available using only analytic methods. Further considerations about the missing states and the full solution of the many anyon problem can be obtained using semiclassical (Illuminati *et al.* 1992; Sporre *et al.* 1991c), numerical (Sporre *et al.* 1991a; Murphy *et al.* 1991; Sporre *et al.* 1991b; Chin and Hu 1992) or perturbative (Chou 1991b; McCabe and Ouvry 1991; Comtet *et al.* 1991; Khare and McCabe 1991; Dasnières de Veigy and Ouvry 1991; Chou *et al.* 1992) methods as we will see in Section 4.

Let us now return to our exact analysis. Under our assumptions, we are now going to show that for $\nu \neq 0, 1$ either all b_{IJ} or all a_{IJ} vanish, if one requires that $\hat{\psi}$ be an eigenfunction of the Hamiltonian $\hat{H}$. The most singular term in $\hat{H}\hat{\psi}$ comes from the Laplacian $\sum_I \partial_I \bar{\partial}_I$ and is explicitly given by

$$-\frac{2}{m}\sum_{I=1}^{N}\left[\sum_{J\neq I}\frac{a_{IJ}}{z_{IJ}}\sum_{K\neq I}\frac{b_{IK}}{\bar{z}_{IK}}\right]\prod_{I<J}(z_{IJ})^{a_{IJ}}(\bar{z}_{IJ})^{b_{IJ}}\phi(z_1,\ldots,z_N;\bar{z}_1,\ldots,\bar{z}_N)+\text{symm} \tag{6.1.22}$$

where for $I > J$ we define $a_{IJ} = a_{JI}$ and $b_{IJ} = b_{JI}$. Of course to satisfy the eigenvalue equation (6.1.16a), this term must cancel. Let us suppose that the set $\mathcal{A}$ of non-integer a_{IJ}'s is not empty and let us call $a_{\alpha\beta}$ an element of $\mathcal{A}$ such that

$$a_{\alpha\beta} \leq a_{IJ} \ , \qquad \forall a_{IJ} \in \mathcal{A} \ . \tag{6.1.23}$$

A moment thought reveals that because of (6.1.23), symmetrization cannot produce accidental cancellations and therefore (6.1.16a) implies that the expression in square brackets in (6.1.22) must vanish, that is

$$b_{\alpha I} = b_{\beta J} = 0 \qquad \forall I \neq \alpha \ , \quad \forall J \neq \beta \ .$$

Upon inserting this into (6.1.19), one gets

$$a_{\alpha I} = \nu \ \ [\text{mod}\, 2] \ , \qquad a_{\beta J} = \nu \ \ [\text{mod}\, 2] \ .$$

The argument can be repeated, until proving that all b_{IJ} vanish and all $a_{IJ} = \nu \ [\text{mod}\, 2]$.

Since the regularity of the wavefunction at coincident points excludes the possibility that some a_{IJ} be equal to $\nu - 2$, $\nu - 4$ and so on for $\nu \in [0, 2)$, we find that the wavefunction (6.1.18) is actually of the form

$$\hat{\psi}_{\text{I}} \equiv \prod_{I<J}(z_{IJ})^{\nu}\, P(z_1,\ldots,z_N;\bar{z}_1,\ldots\bar{z}_N) \ , \tag{6.1.24a}$$

where possible extra powers coming from $a_{IJ} = \nu{+}2, \nu{+}4, \ldots$, have been absorbed in P and the same has been done for symmetrization. Hence P is completely symmetric under any permutation of the particle labels.

If one starts by supposing that some b_{IJ} is not an integer, the proof is exactly as above, and one finds that $\hat{\psi}$ must have the following expression

$$\hat{\psi}_{\rm II} \equiv \prod_{I<J} (\bar{z}_{IJ})^{2-\nu} P(z_1, \ldots, z_N; \bar{z}_1, \ldots, \bar{z}_N) \quad , \tag{6.1.24b}$$

where, once again, the regularity for $z_{IJ} \to 0$ has been required. We will call the states in (6.1.24a) and (6.1.24b) Type-I and Type-II states respectively to emphasize the different structure of their multivalued prefactors.

We remark that for $\nu \neq 0,1$ the states $\hat{\psi}_{\rm I}$ and $\hat{\psi}_{\rm II}$ do not overlap and form two distinct families. In fact the radial behavior of the function P in (6.1.23) at points of coincidence is $|z_{IJ}|^{\ell}$ with ℓ even and non-negative. This prevents transforming Type-I states into Type-II states by writing for instance

$$P = \prod_{I<J} (\bar{z}_{IJ})^2 \, |z_{IJ}|^{-2\nu} \, \tilde{P} \quad . \tag{6.1.25}$$

On the contrary $\hat{\psi}_{\rm I}$ and $\hat{\psi}_{\rm II}$ states do overlap for $\nu = 0,1$. It is important to realize that for $\nu = 0$, *all* bosonic wavefunctions can be written as in (6.1.23) with P a polynomial in $z_I, \bar{z}_I$. Also *all* fermionic ($\nu = 1$) wavefunctions can be written in the representation (6.1.23): in this case however P is not necessarily a polynomial in $z_I, \bar{z}_I$, as is clear from the following example. Consider three fermions described by the wavefunction

$$\hat{\psi} = z_{12}\bar{z}_3 + z_{23}\bar{z}_1 + z_{31}\bar{z}_2 \quad . \tag{6.1.26}$$

This is an eigenfunction both of energy and angular momentum, that it is completely antisymmetric under particle exchanges and that it can be written as in (6.1.24a) with

$$P = \frac{\bar{z}_3}{z_{13}z_{23}} + \frac{\bar{z}_1}{z_{12}z_{13}} + \frac{\bar{z}_2}{z_{12}z_{32}} \quad , \tag{6.1.27}$$

which is finite at all coincidence points but is not a polynomial [14]. The special case in which P is a polynomial also for fermions is when all particles are in the first Landau level (Dunne *et al.* 1991a).

In the next sections we are going to study in detail the eigenvalue problem (6.1.17) with wavefunctions $\hat{\psi}$ of the form (6.1.23). However already at this point, it is clear that the simplest of all such wavefunctions are obtained by setting $P = 1$, namely

$$\hat{\psi}_{\rm I}^{(0)} = \prod_{I<J} (z_{IJ})^{\nu} \tag{6.1.28a}$$

for Type-I, and

$$\hat{\psi}_{\rm II}^{(0)} = \prod_{I<J} (\bar{z}_{IJ})^{2-\nu} \tag{6.1.28b}$$

for Type-II. They are energy and angular momentum eigenstates with eigenvalues

$$\begin{aligned} E_{\rm I}^{(0)} &= \omega \, \frac{N}{2} \quad , \\ J_{\rm I}^{(0)} &= \frac{\nu}{2} N(N-1) \quad , \end{aligned} \tag{6.1.29a}$$

[14] Moreover, its value in $z_{12} \to 0$ for instance, depends on $\arg(z_{12})$.

and

$$E_{\rm II}^{(0)} = \omega \left(\frac{2-\nu}{2} N(N-1) + \frac{N}{2} \right) \; , \qquad J_{\rm II}^{(0)} = -\frac{2-\nu}{2} N(N-1) \; , \tag{6.1.29b}$$

respectively. Therefore for $\nu \neq 0, 1$ $\hat{\psi}_{\rm I}^{(0)}$ and $\hat{\psi}_{\rm II}^{(0)}$ form two distinct and preferred "base" states upon which one can build two families (Type-I and Type-II) of regular functions as we will see in the next section.

6.2 Step Operators and Wavefunctions

A possible strategy to construct exact many-body wavefunctions for anyons in a magnetic field is to start from the "base" states

$$\hat{\psi}_{\rm I}^{(0)} = \prod_{I<J} (z_{IJ})^{\nu} \tag{6.2.1}$$

for Type-I, and

$$\hat{\psi}_{\rm II}^{(0)} = \prod_{I<J} (\bar{z}_{IJ})^{2-\nu} \tag{6.2.2}$$

for Type-II, and act on them step operators which increase the energy or the angular momentum. For simplicity and to avoid repetitions, we limit our discussion to Type-I wavefunctions; Type-II wavefunctions can be treated in the same way with $z_I \to \bar{z}_I$, $\nu \to 2-\nu$.

To construct the appropriate step operators, let us observe that the Hamiltonian $\hat{H}$ in (6.2.17) and the angular momentum $J_{\rm c}$ in (6.2.10) can be written as

$$\hat{H} = \omega \sum_{I=1}^{N} a_I^{\dagger} a_I \; , \tag{6.2.3a}$$

$$J_{\rm c} = \sum_{I=1}^{N} \left(b_I^{\dagger} b_I - a_I^{\dagger} a_I \right) \; , \tag{6.2.3b}$$

where the operators

$$a_I^{\dagger} = \bar{z}_I - \frac{2}{eB} \partial_I \; , \qquad a_I = \bar{\partial}_I \; , \tag{6.2.4a}$$

$$b_I^{\dagger} = z_I - \frac{2}{eB} \bar{\partial}_I \; , \qquad b_I = \partial_I \; , \tag{6.2.4b}$$

satisfy $[a_I, a_J^{\dagger}] = [b_I, b_J^{\dagger}] = \delta_{IJ}$, all other possible commutators being zero.

From (6.2.3) and (6.2.4) we see that the $a_I^{\dagger}$ and $b_I^{\dagger}$ act as "creation" operators while the a_I and b_I act as "annihilation" operators (this will be made more precise below). Therefore our general strategy will be to construct multi-anyon wavefunctions by acting with these creation operators on the base states in (6.2.1) and (6.2.2). We must, however, respect the statistics and hard-core requirements for the resulting wavefunctions as discussed in the previous section. To respect the statistics we may only use *symmetric* (in the particle labels I) combinations of the step operators — the statistics being encoded entirely in the base states (6.2.1) and (6.2.2). To respect the hard-core requirement we ask that wavefunctions vanish at coincident points (for $\nu \neq 0$). Therefore we will have reject those combinations of step operators which produce singular (at coincident points) wavefunctions.

The reason we called a_I annihilation operator (for Type-I states) is that it vanishes on the "base" state, namely

$$a_I \, \hat{\psi}_{\rm I}^{(0)} = 0 \tag{6.2.5}$$

for all I. Moreover, from (6.2.3a) it is immediate to see that

$$\hat{H} \, \hat{\psi}_{\rm I}^{(0)} = 0 \ . \tag{6.2.6}$$

Also b_I is somehow an annihilation operator, but its action on the base state is not so simple; indeed one finds

$$b_I \, \hat{\psi}_{\rm I}^{(0)} = \left(\sum_{J \neq I} \frac{\nu}{z_I - z_J} \right) \hat{\psi}_{\rm I}^{(0)} \ . \tag{6.2.7}$$

However, the symmetric combination $\sum_{I=1}^{N} b_I$ does annihilate $\hat{\psi}_{\rm I}^{(0)}$, *i.e.*

$$\sum_{I=1}^{N} b_I \, \hat{\psi}_{\rm I}^{(0)} = 0 \ , \tag{6.2.8}$$

and this justifies our assertion. When $\nu \neq 0, 1$ (*i.e.* for generic anyons), higher powers of b_I (as well as products of $a_I^\dagger$ and b_I) produce singularities when acting on $\hat{\psi}_{\rm I}^{(0)}$, which cannot be removed by symmetrization. Thus these operators cannot be used and only the "creation" modes ($a_I^\dagger$ and $b_I^\dagger$) have to be considered. When ν is integer ($\nu = 0, 1$) the situation is different. In fact, for $\nu = 0$ the operator $(b_I)^n$ annihilates $\hat{\psi}_{\rm I}^{(0)}$ and for $\nu = 1$ it produces regular functions when acting on $\hat{\psi}_{\rm I}^{(0)}$. This property is what singles out the integer (0 or 1) values of the statistics parameter ν. In conclusion, to produce acceptable wavefunctions from the base state (6.2.1) for non-integer ν, we can use only the creation operators $a_I^\dagger$ and $b_I^\dagger$, while for $\nu = 1$ we may also use b_I [15]. In particular one may show that, for $\nu = 1$, symmetric combinations of $a_I^\dagger$, $b_I^\dagger$ and b_I acting on (6.2.1) produce *all* fermionic

[15]Note that also the combinations $b_I^{\dagger n} b_I$ (appropriately symmetrized) produce regular wavefunctions when acting on $\hat{\psi}_{\rm I}^{(0)}$. However, these have the form of polynomials multiplying $\hat{\psi}_{\rm I}^{(0)}$ and can also be obtained by the action of $b_I^\dagger$'s alone. Therefore the combinations $b_I^{\dagger n} b_I$ are redundant.

states. We can give an easy example of this fact. Let us consider the lowest angular momentum state with two fermions in the ground level and one in the first excited level which is described by the wavefunction (6.1.26). Some very simple algebra reveals that

$$\hat{\psi} = z_{12}\bar{z}_3 + z_{23}\bar{z}_1 + z_{31}\bar{z}_2 = \frac{1}{2}\sum_{I=1}^{3} a_I^\dagger b_I^2 \, \hat{\psi}_{\rm I}^{(0)} \quad .$$

Notice that there is no analogue state involving the b_I's for $\nu \neq 1$, since the corresponding wavefunction would be singular.

To treat the problem systematically, let us recall that a basis in the ring of symmetric polynomials in $2N$ variables ($a_I^\dagger$ and $b_I^\dagger$ with $I = 1, ..., N$) is given by

$$C_{\ell n} = \sum_{I=1}^{N} a_I^{\dagger \ell} b_I^{\dagger n} \quad , \tag{6.2.9}$$

where ℓ, n are non-negative integers such that $\ell + n \leq N$. In particular the operators $C_{10} = \sum_{I=1}^{N} a_I^\dagger$ and $C_{01} = \sum_{I=1}^{N} b_I^\dagger$ describe completely the center of mass excitations and were originally introduced in (Johnson and Canright 1990). More generally, the operators $C_{\ell n}$, which respect the statistics properties of the base state $\hat{\psi}_{\rm I}^{(0)}$, are step operators in energy and angular momentum, in fact

$$\left[\hat{H}, C_{\ell n}\right] = \ell\omega \, C_{\ell n} \quad , \tag{6.2.10a}$$

$$[J, C_{\ell n}] = (n - \ell) \, C_{\ell n} \quad . \tag{6.2.10b}$$

Therefore $C_{\ell n}$ can be used on the base state to produce exact energy and angular momentum anyonic eigenstates (Dunne *et al.* 1992a; Grundberg *et al.* 1991b). However, since in general they contain derivatives, we are not guaranteed that this procedure produces *regular* eigenstates and particular care must be used to identify which particular $C_{\ell n}$'s produce acceptable states. In any case it is clear that our procedure will lead to states of the form (6.1.24a) with $P(z_1, \ldots z_N, \bar{z}_1, \ldots z_N)$ being a polynomial function. Thus, for $\nu \neq 1$ it is not guaranteed that the $C_{\ell n}$'s produce *all* anyonic eigenstates.

To select the "safe" operators, let us begin with $C_{0n} = \sum_{I=1}^{N} b_I^{\dagger n}$ (with $n \leq N$). Using (6.2.4), we obtain

$$C_{0n} \, \hat{\psi}_{\rm I}^{(0)} = \left(\sum_I z_I^n\right) \hat{\psi}_{\rm I}^{(0)} \quad , \tag{6.2.11}$$

which is regular at all coincident points. This shows that all operators C_{0n}'s with $n \leq N$ can be safely applied to the base state to generate regular wavefunctions, which have the following form

$$\hat{\psi}(\{\lambda_{0n}\}) = \prod_{n=0}^{N} (C_{0n})^{\lambda_{0n}} \, \hat{\psi}_{\rm I}^{(0)} \tag{6.2.12}$$

for arbitrary non-negative integers λ_{0n}.

From (6.2.9a) and (6.2.6) it is immediate to realize that the functions (6.2.11) are annihilated by the Hamiltonian $\hat{H}$ so that their energy is $E = N\omega/2$. They correspond to configurations in which all particles are in the lowest Landau level with total angular momentum

$$J(\{\lambda_{0n}\}) = \nu\frac{N(N-1)}{2} + \sum_{n=0}^{N} n\lambda_{0n} \ . \tag{6.2.13}$$

In order to go to excited energy levels, we have to consider necessarily operators $C_{\ell n}$ with $\ell \geq 1$; since these objects contain derivatives with respect to z_I, they can lead to singular wavefunctions and a careful study is needed to select those combinations which preserve regularity. In this respect an asymmetry between $0 < \nu < 1$ and $1 < \nu < 2$ seems to appear: one can surely apply one derivative ∂_I to $\hat{\psi}_{\rm I}^{(0)}$ for $1 < \nu < 2$ without getting a singular function for $z_{IJ} \to 0$; the same seems not to be true for $0 < \nu < 1$. On the contrary, we will show that actually there is no asymmetry and that C_{1m} (with $m = 0, 1, \ldots$) applied to $\hat{\psi}_{\rm I}^{(0)}$ still gives a regular function for $z_{IJ} \to 0$ for $0 < \nu < 2$. In fact, from (6.2.6) and (6.2.4),

$$\begin{aligned} C_{1m}\hat{\psi}_{\rm I}^{(0)} &= \sum_I \left(\bar{z}_I - \partial_I\right)\left(z_I - \bar{\partial}_I\right)^m \hat{\psi}_{\rm I}^{(0)} \\ &= \sum_I \left(\bar{z}_I - \partial_I\right) z_I^m \hat{\psi}_{\rm I}^{(0)} \\ &= -\sum_I z_I^m \partial_I \hat{\psi}_{\rm I}^{(0)} + \sum_I \left(\bar{z}_I z_I^m - m z_I^{m-1}\right)\hat{\psi}_{\rm I}^{(0)} \ , \end{aligned} \tag{6.2.14}$$

where for simplicity we have set $eB = 2$. Only the first term is potentially singular; however

$$\sum_I z_I^m \,\partial_I \hat{\psi}_{\rm I}^{(0)} = \sum_I z_I^m \sum_{J \neq I} \frac{\nu}{z_I - z_J}\,\hat{\psi}_{\rm I}^{(0)} = \frac{1}{2}\nu \sum_{I \neq J} \frac{z_I^m - z_J^m}{z_I - z_J}\,\hat{\psi}_{\rm I}^{(0)} \tag{6.2.15}$$

is regular because the factor in front of $\hat{\psi}_{\rm I}^{(0)}$ is a polynomial.

Actually, C_{1m} can be repeatedly applied to $\hat{\psi}_{\rm I}^{(0)}$ (or to $\hat{\psi}(\{\lambda_{0n}\})$) without generating singularities. In fact, let us consider the state

$$C_{1m_1} C_{1m_2} \cdots \hat{\psi}_{\rm I}^{(0)} \ ,$$

and write C_{1m_i} as

$$C_{1m_i} = \sum_{I=1}^{N} \left(\bar{z}_I - \partial_I\right)\left(z_I - a_I\right)^{m_i} \ .$$

The destruction operators a_I either annihilate $\hat{\psi}_{\rm I}^{(0)}$ or cancel one of the $a_I^\dagger$ in the C_{1m_j} $(j > i)$. Therefore the potentially singular terms are again produced by

$\sum_{I=1}^{N} z_I^n \partial_I$ $(0 \le n \le m_i)$ for which (6.2.15) holds. Therefore we have enlarged our family of allowed states to include also

$$\begin{aligned} \hat{\psi}_{\rm I}\left(\{\lambda_{1m}, \lambda_{0\ell}\}\right) &= \prod_{m=0}^{N-1} (C_{1m})^{\lambda_{1m}} \prod_{\ell=0}^{N} (C_{0\ell})^{\lambda_{0\ell}} \hat{\psi}_{\rm I}^{(0)} \\ &= \prod_{n=0}^{1} \prod_{m=0}^{N-n} (C_{nm})^{\lambda_{nm}} \prod_{I<J} (z_{IJ})^{\nu} \ , \end{aligned} \tag{6.2.16a}$$

where λ_{nm} are non-negative integers, with energy and angular momentum given by

$$E = \omega \left(\sum_{m=0}^{N-1} \lambda_{1m} + \frac{N}{2} \right) \tag{6.2.16b}$$

$$J = \nu \frac{N(N-1)}{2} + \sum_{n=0}^{1} \sum_{m=0}^{N-n} (m-n)\lambda_{nm} \ . \tag{6.2.16c}$$

The family (6.2.16a) can be further enlarged by observing that some of the states (6.2.12) are of the form

$$\hat{\psi}_{\rm I}^{(s)} = \prod_{I<J} (z_{IJ})^{\nu+2s} \ , \qquad s = 1, 2, \ldots \ .$$

Thus a string of operators $C_{n_1 m_1} C_{n_2 m_2} \cdots C_{n_i m_i}$ with $\sum_{j=1}^{i} n_j \le 2s$ can be applied to $\hat{\psi}_{\rm I}^{(s)}$ without generating singularities, because such a string contains at most derivatives of order $\sum_{j=1}^{i} n_j$ with respect to z_I.

Moreover, the same argument of (6.2.15) can be easily extended to prove that $\sum_{I=1}^{N} z_I^m \partial_I^{2k+1}$ applied to $\prod_{I<J} (z_{IJ})^{\rho}$ gives $\prod_{I<J} (z_{IJ})^{\rho-2k}$ times a polynomial (notice that the exponent of $\prod_{I<J} (z_{IJ})^{\rho}$ is lowered by $2k$ and not by $2k+1$ as one may naively expect). By the same reasoning that led to the states (6.2.16a), we can write the following large family of wavefunctions

$$\hat{\psi}_{\rm I}(\{\lambda_{nm}\}) = \prod_{n=0}^{N} \prod_{m=0}^{N-n} (C_{nm})^{\lambda_{nm}} \prod_{I<J} (z_{IJ})^{\nu+2s} \ , \qquad s = 0, 1, 2, \ldots \tag{6.2.17}$$

where the non-negative integers λ_{nm} are constrained by

$$\sum_{n=0}^{N} \sum_{m=0}^{N-n} \{n\} \lambda_{nm} = 2s \tag{6.2.18}$$

where $\{n\}$ is the largest even number not greater than n. The energy and angular momentum of the states (6.2.17) are

$$E = \omega\left(\sum_{n=1}^{N}\sum_{m=0}^{N-n} n\lambda_{nm} + \frac{N}{2}\right) \tag{6.2.19a}$$

$$J = (\nu + 2s)\frac{N(N-1)}{2} + \sum_{n=0}^{N}\sum_{m=0}^{N-n}(m-n)\lambda_{nm} \quad . \tag{6.2.19b}$$

In the constraint (6.2.18) we do not allow the left-hand side to be less than $2s$ because the corresponding states are already contained in (6.2.17) with s replaced by $s-1$, $s-2, \ldots$. We remark that the states (6.2.17) are not all linearly independent because of highly non-trivial polynomial relations between products of the C_{nm}'s (Dunne *et al.* 1992a). It is also interesting to note that the left-hand side of (6.2.18) does not depend on n, but only on $\{n\}$ and therefore it jumps by two steps; this confirms that there is no asymmetry between the cases $0 < \nu < 1$ and $1 < \nu < 2$.

A similar discussion can be repeated starting from $\hat{\psi}_{\mathrm{II}}^{(0)}$, leading to the Type-II states

$$\hat{\psi}_{\mathrm{II}}(\{\lambda_{nm}\}) = \prod_{n=0}^{N}\prod_{m=0}^{N-n}(C_{nm})^{\lambda_{nm}}\prod_{I<J}(\bar{z}_{IJ})^{2-\nu+2s} \quad , \qquad s = 0, 1, 2, \ldots \tag{6.2.20}$$

where the non-negative integers λ_{nm} are constrained by

$$\sum_{n=0}^{N}\sum_{m=0}^{N-n}\{m\}\lambda_{nm} = 2s \quad . \tag{6.2.21}$$

The energy and angular momentum are given by

$$E = \omega\left(\sum_{n=1}^{N}\sum_{m=0}^{N-n} n\lambda_{nm} + \frac{N}{2} + (2-\nu+2s)\frac{N(N-1)}{2}\right) \quad , \tag{6.2.22a}$$

$$j = -(2-\nu+2s)\frac{N(N-1)}{2} + \sum_{n=0}^{N}\sum_{m=0}^{N-n}(m-n)\lambda_{nm} \quad . \tag{6.2.22b}$$

So far we have considered symmetric polynomials in the operators $a_I^\dagger$ and $b_I^\dagger$ constructed from the symmetric step operators C_{nm}. However there is another way of obtaining symmetric polynomials in $a_I^\dagger$ and $b_I^\dagger$, namely multiplying an even number of antisymmetric polynomials. The latter can be written as determinants in $a_I^\dagger$ and $b_I^\dagger$, but as we have already stressed, not all such expressions produce regular wavefunctions and are allowed. A detailed analysis is therefore necessary to select "safe" operators, and in this way a further set of regular wavefunctions can be constructed. We refer the reader to the original literature (Dunne *et al.* 1992a) for a complete discussion of this procedure. Here instead we work out explicitly the two-anyon case starting from the general states (6.2.17) and (6.2.20). Eq. (6.2.17) for $s = 0$ gives the wavefunctions

$$\hat{\psi}_{\rm I}\left(\{\lambda_{nm}\}\right) = (C_{10})^{\lambda_{10}}(C_{01})^{\lambda_{01}}(C_{11})^{\lambda_{11}}(C_{02})^{\lambda_{02}}(z_{12})^{\nu} \ . \tag{6.2.23}$$

The center of mass dependence can be easily factorized by defining,

$$\begin{aligned} A^{\dagger} &= a_1^{\dagger} + a_2^{\dagger} \ , & B^{\dagger} &= b_1^{\dagger} + b_2^{\dagger} \ , \\ a^{\dagger} &= a_1^{\dagger} - a_2^{\dagger} \ , & b^{\dagger} &= b_1^{\dagger} - b_2^{\dagger} \ , \end{aligned} \tag{6.2.24}$$

so that we can write

$$C_{10} = A^{\dagger} \ ; \quad C_{01} = B^{\dagger} \ ; \quad C_{11} = \frac{1}{2}\left(A^{\dagger}B^{\dagger} + a^{\dagger}b^{\dagger}\right) \ ; \quad C_{02} = \frac{1}{2}\left(B^{\dagger 2} + b^{\dagger 2}\right) \ . \tag{6.2.25}$$

Using this notation, one can write a more convenient basis in the space spanned by the states (6.2.23), namely

$$\hat{\psi}_{\rm I}(N,M,n,m) = A^{\dagger N}B^{\dagger M}a^{\dagger n}b^{\dagger m}(z_{12})^{\nu} \tag{6.2.26}$$

where N, M, n, m are non-negative integers, and $m - n = 0, 2, 4, \ldots$

Eq. (6.2.26) reproduces half of the well-known eigenfunctions for two anyons in a constant magnetic field. Indeed, after some elementary algebra we find

$$\hat{\psi}_{\rm I}(N,M,n,m) \propto Z^{M-N}L_N^{M-N}\left(\frac{eB}{2}|Z|^2\right)(z_{12})^{m-n+\nu}L_n^{m-n+\nu}\left(\frac{eB}{2}|z_{12}|^2\right) \tag{6.2.27}$$

where $Z = \frac{1}{\sqrt{2}}(z_1 + z_2)$ is proportional to the center of mass coordinate. The other eigenfunctions come from the Type-II states (6.2.20),

$$\hat{\psi}_{\rm II}\left(\{\lambda_{nm}\}\right) = (C_{10})^{\lambda_{10}}(C_{01})^{\lambda_{01}}(C_{11})^{\lambda_{11}}(C_{20})^{\lambda_{20}}(\bar{z}_{12})^{2-\nu} \ . \tag{6.2.28}$$

Using the definitions (6.2.25) and

$$C_{20} = \frac{1}{2}\left(A^{\dagger 2} + a^{\dagger 2}\right) \ , \tag{6.2.29}$$

these may be written as

$$\begin{aligned} \hat{\psi}_{\rm II}(N,M,n,m) &= A^{\dagger N}B^{\dagger M}a^{\dagger n}b^{\dagger m}(\bar{z}_{12})^{2-\nu} \\ &\propto Z^{M-N}L_N^{M-N}\left(\frac{eB}{2}|Z|^2\right)(\bar{z}_{12})^{n-m+2-\nu}L_m^{n-m+2-\nu}\left(\frac{eB}{2}|\bar{z}_{12}|^2\right) \end{aligned} \tag{6.2.30}$$

for $n - m = 0, 2, \ldots$

It is important to realize that in the particular case of two particles both the bosonic limit $\nu \to 0$ and the fermionic limit $\nu \to 1$ are smooth, so that the wavefunctions in (6.2.27) and (6.2.30) are the *full* solution of the two-anyon problem. In particular all the two-fermion wavefunctions, which are [16]

[16] We write only the factors depending on the relative coordinate and understand those depending on the center of mass coordinate.

$$(z_{12})^{m-n+1} L_n^{m-n+1}\left(\frac{eB}{2}|z_{12}|^2\right) \tag{6.2.31a}$$

with m and n non-negative integers such that $m - n = 0, 2, \ldots$, and

$$(\bar{z}_{12})^{n-m+1} L_m^{n-m+1}\left(\frac{eB}{2}|z_{12}|^2\right) \propto (z_{12})^{m-n-1} L_{n+1}^{m-n-1}\left(\frac{eB}{2}|z_{12}|^2\right) \tag{6.2.31b}$$

for $m - n = 0, -2, \ldots$, are reproduced by taking $\nu = 1$ in (6.2.27) and (6.2.30).

Furthermore, the whole spectrum of states for two anyons is given by (6.2.17) and (6.2.20) with $s = 0$. In fact for $s = 1$ we could for example also consider the state

$$C_{20}(z_{12})^{2+\nu} = C_{20}\left[\det\begin{pmatrix} 1 & 1 \\ b_1^\dagger & b_2^\dagger \end{pmatrix}\right]^2 (z_{12})^\nu \ . \tag{6.2.32}$$

However, from the identities

$$\det\begin{pmatrix} 1 & 1 \\ a_1^\dagger & a_2^\dagger \end{pmatrix}^2 = \det\left[\begin{pmatrix} 1 & 1 \\ a_1^\dagger & a_2^\dagger \end{pmatrix}\begin{pmatrix} 1 & a_1^\dagger \\ 1 & a_2^\dagger \end{pmatrix}\right] = \det\begin{pmatrix} 2 & C_{10} \\ C_{10} & C_{20} \end{pmatrix} ,$$

$$\det\begin{pmatrix} 1 & 1 \\ b_1^\dagger & b_2^\dagger \end{pmatrix}^2 = \det\begin{pmatrix} 2 & C_{01} \\ C_{01} & C_{02} \end{pmatrix} ,$$

$$\det\left[\begin{pmatrix} 1 & 1 \\ a_1^\dagger & a_2^\dagger \end{pmatrix}\begin{pmatrix} 1 & b_1^\dagger \\ 1 & b_2^\dagger \end{pmatrix}\right] = \det\begin{pmatrix} 2 & C_{01} \\ C_{10} & C_{11} \end{pmatrix} ,$$

it is easy to prove that [17]

$$2C_{20}\left[\det\begin{pmatrix} 1 & 1 \\ b_1^\dagger & b_2^\dagger \end{pmatrix}\right]^2 = C_{10}^2\left(2C_{02} - C_{01}^2\right) + \left(2C_{11} - C_{01}C_{10}\right)^2 \ . \tag{6.2.33}$$

This shows that the state (6.2.32) is a linear combination of the states (6.2.23). In general we see that among the independent symmetric polynomials C_{nm} relevant for $N = 2$, the only one not appearing in (6.2.23) is C_{20}. However, from (6.2.17) and (6.2.18) it is clear that for any value of s, C_{20} always appears in the combination (6.2.33) not to produce singularities. Thus, actually no new state is obtained for $s > 0$. Identities like (6.2.33) indicate that the states (6.2.17) are not all independent of each other. Unfortunately, such identities are not easy to derive for $N \geq 3$; in (Dunne *et al.* 1992a) the case $N = 3$ has been considered in detail, but the general case is still an open problem.

[17] Eq. (6.2.33) does not contradict the fact that the C_{nm}'s $(m + n \leq N)$ are a minimal basis in the ring of symmetric polynomials; in general polynomial relations among these objects can be written, but it is not possible to solve them in order to express *one* C_{nm} $(m + n \leq N)$ as a polynomial in the others.

6.3 Closed-Form Eigenstates

In this section we solve directly the eigenstate differential equations (6.1.16) to find solutions in a closed-form. These are of course contained within the states discussed in the previous section, but for the sake of clarity here we present their explicit functional form. This approach also illustrates the separation of the dependence on the center of mass coordinate in the N-body anyon problem and yields states which are the N-body generalization of those found previously by Y.-S. Wu (Wu 1984b) in the three-body problem for the related system of anyons in a harmonic potential [18].

Our starting point is the "reduced" Hamiltonian $\hat{H}$ and the corresponding "reduced" wave function $\hat{\psi}$ defined in (6.1.17). We shall seek Type-I [19] eigenstates $\hat{\psi}_{\rm I}$ of $\hat{H}$ and of the angular momentum J_c of the form (6.1.24), *i.e.*

$$\hat{\psi}_{\rm I} = \prod_{I<J} (z_{IJ})^{\nu}\; P\left(z_1, \ldots z_N; \bar{z}_1, \ldots \bar{z}_N\right) \quad , \tag{6.3.1}$$

so that the eigenstate conditions (6.1.16) on $\hat{\psi}_{\rm I}$ become eigenstate conditions on P. Indeed, for $\hat{\psi}_{\rm I}$ to be an eigenstate of J_c, we must require that P be also an eigenstate of the same operator J_c, since the statistical prefactor $\prod\limits_{I<J}^{N} (z_{IJ})^{\nu}$ is itself an eigenstate of J_c, with eigenvalue $\nu N(N-1)/2$. Instead, for $\hat{\psi}_{\rm I}$ to be an eigenstate of $\hat{H}$, we see that P must satisfy a modified differential equation, which reads

$$\sum_{I=1}^{N} \left(-\frac{2}{m}\partial_I + \omega \bar{z}_I\right) \bar{\partial}_I P - \frac{2}{m}\nu \sum_{I<J} \left(\frac{\bar{\partial}_I - \bar{\partial}_J}{z_I - z_J}\right) P = \left(E - \omega\frac{N}{2}\right) P \ . \tag{6.3.2}$$

The general solution of this partial differential equation is not known, but some families of solutions can be found quite easily. For example we can make the following *Ansatz* (Dunne *et al.* 1991a)

$$P\left(z_1, \ldots, z_N; \bar{z}_1, \ldots, \bar{z}_N\right) = \mathcal{S}^{(d_+)}\left(z_1, \ldots, z_N\right) \phi\left(\frac{eB}{2}\sum_{I=1}^{N} |z_I|^2\right) \quad , \tag{6.3.3}$$

where $\mathcal{S}^{(d_+)}(z_1, \ldots, z_N)$ is a totally symmetric homogeneous analytic polynomial of degree d_+ in z_I, and ϕ is a function to be determined. Inserting (6.3.3) into (6.3.2) we obtain the following differential equation for ϕ

---

[18]Note that the harmonic potential problem involves the Hamiltonian (6.1.9) without the angular momentum term. But since H and J_c commute, the eigenvalue problems are essentially the same.

[19]The steps discussed below for Type-I states may be repeated similarly for Type-II states, leading to analogous equations whose solutions are listed in (6.3.6b) and (6.3.37b).

$$y\,\phi''(y) + \left(d_+ + N + \frac{\nu}{2}N(N-1) - y\right)\phi'(y) + \left(\frac{E}{\omega} - \frac{N}{2}\right)\phi(y) = 0 \qquad (6.3.4)$$

where $y = (eB/2)\sum_{I=1}^{N}|z_I|^2$ and the primes denote derivatives with respect to y. This is a differential equation of generalized Laguerre type (see Appendix A), with solution

$$\phi(y) = L_n^{d_+ + N - 1 + \frac{\nu}{2}N(N-1)}(y) \ , \qquad (6.3.5)$$

where $n = \frac{E}{\omega} - \frac{N}{2}$ is a non-negative integer.

Thus, restoring the exponential factor, our *Ansatz* (6.3.3) yields the following exact N-anyon Type-I wavefunctions

$$\psi_{\mathrm{I}} = \prod_{I<J}(z_{IJ})^{\nu}\mathcal{S}^{(d_+)}(z_1,\ldots,z_N)$$
$$\times L_n^{d_+ + N - 1 + \frac{\nu}{2}N(N-1)}\left(\frac{eB}{2}\sum_{I=1}^{N}|z_I|^2\right)\exp\left(-\frac{eB}{4}\sum_{I=1}^{N}|z_I|^2\right) \ ,$$

with energy and angular momentum given by

$$\begin{aligned} E_{\mathrm{I}} &= \omega\left(n + \frac{N}{2}\right) \ , \qquad && n = 0,1,2,\ldots, \ ; \\ J_{\mathrm{I}} &= \frac{\nu}{2}N(N-1) + d_+ \ , \qquad && d_+ = 0,1,2,\ldots \ , \end{aligned} \qquad (6.3.6a)$$

An analogous *Ansatz* for Type-II states leads to

$$\psi_{\mathrm{II}} = \prod_{I<J}(\bar{z}_{IJ})^{2-\nu}\mathcal{S}^{(d_+)}(\bar{z}_1,\ldots,\bar{z}_N)$$
$$\times L_n^{d_- + \frac{2-\nu}{2}N(N-1) + N - 1}\left(\frac{eB}{2}\sum_{I=1}^{N}|z_I|^2\right)\exp\left(-\frac{eB}{4}\sum_{I=1}^{N}|z_I|^2\right) \ ,$$

with

$$\begin{aligned} E_{\mathrm{II}} &= \omega\left(n + \frac{N}{2} + d_- + \frac{2-\nu}{2}N(N-1)\right) \ , \qquad && n = 0,1,\ldots \ , \\ J_{\mathrm{II}} &= -\frac{(2-\nu)}{2}N(N-1) - d_- \ , \qquad && d_- = 0,1,\ldots \ . \end{aligned} \qquad (6.3.6b)$$

More general solutions than those in (6.3.6) can be obtained with a more general *Ansatz* than (6.3.3) for the Type-I states, which may be achieved by factoring out the dependence on the center of mass coordinate proportional to $(z_1+\ldots+z_N)$. Such a factorization is of course quite natural because the "complicated" term $\sum_{I<J}\frac{\bar{\partial}_I - \bar{\partial}_J}{z_I - z_J}$ in (6.3.2) depends only on the *differences* of the coordinates. To accomplish the required separation we first make the following (linear) change of variables

$$z_1, z_1, \ldots, z_N \longrightarrow Z, w_1, w_2, \ldots w_{N-1} \qquad (6.3.7)$$

where

$$
\begin{aligned}
Z &= \frac{1}{\sqrt{N}}\left(z_1 + z_2 + \ldots + z_N\right) \quad , \\
w_1 &= \frac{1}{\sqrt{2}}\left(z_1 - z_2\right) \quad , \\
w_2 &= \frac{1}{\sqrt{6}}\left(z_1 + z_2 - 2z_3\right) \quad , \\
&\vdots \\
w_{N-1} &= \frac{1}{\sqrt{N(N-1)}}\left(z_1 + z_2 + \ldots + z_{N-1} - (N-1)z_N\right) \quad .
\end{aligned}
\tag{6.3.8}
$$

These formulas may be summarized succinctly as

$$
Z = \frac{1}{\sqrt{N}}\sum_{I=1}^{N} z_I \quad , \tag{6.3.9a}
$$

$$
w_i = \sum_{I=1}^{N} u_i^{(I)} z_I \quad , \tag{6.3.9b}
$$

where $\boldsymbol{u}^{(I)}$ are the N vectors (each one with $N-1$ components indexed by i) usually considered to represent the roots and weights of $SU(N)$ (see for example (Georgi 1982)). These vectors, together with their main properties, are listed below

$$
\begin{aligned}
\boldsymbol{u}^{(1)} &= \left(\frac{1}{\sqrt{2}}, \frac{1}{\sqrt{6}}, \frac{1}{\sqrt{12}}, \ldots, \frac{1}{\sqrt{(m+1)(m+2)}}, \ldots, \frac{1}{\sqrt{(N-1)N}}\right) \quad , \\
\boldsymbol{u}^{(2)} &= \left(-\frac{1}{\sqrt{2}}, \frac{1}{\sqrt{6}}, \frac{1}{\sqrt{12}}, \ldots, \frac{1}{\sqrt{(m+1)(m+2)}}, \ldots, \frac{1}{\sqrt{(N-1)N}}\right) \quad , \\
\boldsymbol{u}^{(3)} &= \left(0, -\sqrt{\frac{2}{3}}, \frac{1}{\sqrt{12}}, \ldots, \frac{1}{\sqrt{(m+1)(m+2)}}, \ldots, \frac{1}{\sqrt{(N-1)N}}\right) \quad , \\
&\vdots \\
\boldsymbol{u}^{(m+1)} &= \left(0, 0, \ldots, -\sqrt{\frac{m}{m+1}}, \frac{1}{\sqrt{(m+1)(m+2)}}, \ldots, \frac{1}{\sqrt{(N-1)N}}\right) \quad , \\
&\vdots \\
\boldsymbol{u}^{(N)} &= \left(0, 0, \ldots, 0, \ldots, 0, \ldots, 0, -\sqrt{\frac{N-1}{N}}\right) \quad .
\end{aligned}
\tag{6.3.10}
$$

In terms of these vectors, the positive roots of $SU(N)$ (there are $\frac{1}{2}N(N-1)$ of these) are simply

$$\boldsymbol{r} = \boldsymbol{u}^{(I)} - \boldsymbol{u}^{(J)} \qquad (I < J) \ , \tag{6.3.11}$$

and the fundamental weights of $SU(N)$ (there are $N-1$ of these) are

$$\Lambda_\ell = \sum_{I=1}^{\ell} \boldsymbol{u}^{(I)} \ , \qquad \ell = 1, \dots, N-1 \ . \tag{6.3.12}$$

Furthermore, we have

$$\sum_{I=1}^{N} u_i^{(I)} = 0 \quad \text{for all} \ \ i = 1, \dots, N-1 \ ; \tag{6.3.13}$$

$$\sum_{I=1}^{N} u_i^{(I)} u_j^{(I)} = \delta_{ij} \quad \text{for all} \ \ i, j \ ; \tag{6.3.14}$$

$$\boldsymbol{u}^{(I)} \cdot \boldsymbol{u}^{(J)} = \delta_{IJ} - \frac{1}{N} \ . \tag{6.3.15}$$

The transformations (6.3.9) to the new set of variables $Z, \boldsymbol{w}$ may be inverted using the $\boldsymbol{u}^{(I)}$ vectors according to

$$z_I = \frac{1}{\sqrt{N}} Z + \sum_{i=1}^{N-1} u_i^{(I)} w_i \tag{6.3.16}$$

for $I = 1, \dots N$. Similarly,

$$\partial_I = \frac{1}{\sqrt{N}} \partial_Z + \sum_{i=1}^{N-1} u_i^{(I)} \partial_{w_i} \ . \tag{6.3.17}$$

Using properties (6.3.13) and (6.3.14), it is easy to see that

$$\sum_{I=1}^{n} \partial_I \bar{\partial}_I = \partial_Z \bar{\partial}_Z + \sum_{i=1}^{N-1} \partial_{w_i} \bar{\partial}_{w_i} \ , \tag{6.3.18}$$

and

$$\sum_{I=1}^{N} \bar{z}_I \bar{\partial}_I = \bar{Z} \bar{\partial}_Z + \sum_{i=1}^{N-1} \bar{w}_i \bar{\partial}_{w_i} \ . \tag{6.3.19}$$

Thus, in the new variables, the differential equation for P becomes

$$\left[\left(-\frac{2}{m} \partial_Z + \omega \bar{Z} \right) \bar{\partial}_Z + \sum_{i=1}^{N-1} \left(-\frac{2}{m} \partial_{w_i} + \omega \bar{w}_i \right) \bar{\partial}_{w_i} - \frac{2\nu}{m} \sum_{\boldsymbol{r}>0} \left(\frac{\boldsymbol{r} \cdot \bar{\partial}_{\boldsymbol{w}}}{\boldsymbol{r} \cdot \boldsymbol{w}} \right) \right] P = \left(E - \omega \frac{N}{2} \right) P \ , \tag{6.3.20}$$

where the final term in the operator appearing on the left-hand side involves a sum over all the positive roots $\boldsymbol{r}$ of $SU(N)$. The angular momentum operator also separates according to

$$\begin{aligned} J &= \sum_{I=1}^{N} \left(z_I \partial_I - \bar{z}_I \bar{\partial}_I \right) \\ &= \left(Z\partial_Z - \bar{Z}\bar{\partial}_Z \right) + \sum_{i=1}^{N-1} \left(w_i \partial_{w_i} - \bar{w}_i \bar{\partial}_{w_i} \right) \quad . \end{aligned} \tag{6.3.21}$$

These properties suggest the following separation of variables,

$$P\left(\boldsymbol{z}, \bar{\boldsymbol{z}}\right) = F\left(Z, \bar{Z}\right) G\left(\boldsymbol{w}, \bar{\boldsymbol{w}}\right) \quad . \tag{6.3.22}$$

Requiring P to be an eigenstate of J implies that F in (6.3.22) must be of the form[20]

$$F(Z, \bar{Z}) = Z^a f\left(\frac{eB}{2} |Z|^2 \right) \quad . \tag{6.3.23}$$

With P separated as in (6.3.22) and (6.3.23), the differential equation (6.3.20) requires that the function f satisfy

$$x f'' + (a + 1 - x) f' + \lambda f = 0 \quad , \tag{6.3.24}$$

where λ is an arbitrary positive integer, $x = (eB/2)|Z|^2$ and the primes denote derivatives with respect to x. This is once again the generalized Laguerre differential equation (see Appendix A), giving

$$F(Z, \bar{Z}) = Z^a L_\lambda^a \left(\frac{eB}{2} |Z|^2 \right) \quad . \tag{6.3.25}$$

We remark that a must be an integer to ensure that F be single-valued, and that a may take also *negative* integer values, provided $a \geq -\lambda$. in particular when a is negative, using the special properties of the generalized Laguerre polynomials, we can write

$$Z^{-|a|} L_\lambda^{-|a|} \left(\frac{eB}{2} |Z|^2 \right) \propto \bar{Z}^{|a|} L_{\lambda-|a|}^{|a|} \left(\frac{eB}{2} |Z|^2 \right) \quad .$$

Therefore the *Ansatz* (6.3.23) leads to two types of solutions $Z^{|a|} L_\lambda^{|a|} \left((eB/2)|Z|^2 \right)$ and $\bar{Z}^{|a|} L_{\lambda-|a|}^{|a|} \left((eB/2)|Z|^2 \right)$ which may be succinctly summarized in (6.3.25).

Having solved for the center of mass dependence, we now address the (more complicated!) dependence on the relative coordinates. With F given by (6.3.23) the differential equation (6.3.20) becomes the following differential equation for $G(\boldsymbol{w}, \bar{\boldsymbol{w}})$:

[20]The $\frac{eB}{2}$ in the argument of f is inserted for convenience.

$$\left[\sum_{i=1}^{N-1}\left(-\frac{2}{m}\partial_{w_i}+\omega\bar{w}_i\right)\bar{\partial}_{w_i}-\frac{2\nu}{m}\sum_{r>0}\left(\frac{r\cdot\bar{\partial}w}{r\cdot w}\right)\right]G=\left(E-\omega\left(\frac{N}{2}+\lambda\right)\right)G\ . \tag{6.3.26}$$

Consider the following *Ansatz* for $G(w,\bar{w})$:

$$G(w,\bar{w})=w_1^{d_1}w_2^{d_2}\dots w_{N-1}^{d_{N-1}}\cdot g\left(\frac{eB}{2}\sum_{i=1}^{N-1}|w_i|^2\right)\ . \tag{6.3.27}$$

Here the monomial factor depends only on the w's, not on the $\bar{w}$'s. Inserting this *Ansatz* into (6.3.26) we obtain a differential equation for the function g,

$$y\,g''+\left(N-1+d+\frac{\nu}{2}N(N-1)-y\right)g'+\left(\frac{E}{\omega}-\frac{1}{2}N-\lambda\right)g=0 \tag{6.3.28}$$

where $d\equiv\sum_{i=1}^{N-1}d_i$, and $y=(eB/2)\sum_{i=1}^{N-1}|w_i|^2$. In going from (6.3.26) and (6.3.27) to (6.3.28) we have used the fact that

$$\sum_{r>0}r_i r_j=N\,\delta_{ij}\ . \tag{6.3.29}$$

Hence

$$\frac{1}{2}\sum_{i=1}^{N-1}|w_i|^2=\frac{1}{2N}\sum_{r>0}|r\cdot w|^2\ , \tag{6.3.30}$$

and so

$$\left(\sum_{r>0}\frac{r\cdot\bar{\partial}w}{r\cdot w}\right)g\left(\frac{eB}{2}\sum_{i=1}^{N-1}|w_i|^2\right)=\frac{eB}{4}N(N-1)\ g'\left(\frac{eB}{2}\sum_{i=1}^{N-1}|w_i|^2\right)\ . \tag{6.3.31}$$

The differential equation (6.3.28) is once again of the generalized Laguerre type, and so we find

$$g=L_{E-\frac{1}{2}N-\lambda}^{d+N-2+\frac{\nu}{2}N(N-1)}\left(\frac{eB}{2}\sum_{i=1}^{N-1}|w_i|^2\right)\ . \tag{6.3.32}$$

Thus, with the *Ansatz* (6.3.27) we obtain the result that the factor $G(w,\bar{w})$ in (6.3.22), describing the dependence of P on the relative coordinates, has the form of a monomial in w multiplied by a Laguerre polynomial in $\left((eB/2)\sum_{i=1}^{N-1}|w_i|^2\right)$. However, this function G must also satisfy the appropriate symmetry condition, which is the requirement that P, and hence G, must be totally symmetric in the z_I and $\bar{z}_I$ coordinates. In terms of the new coordinates w, this means that $G(w,\bar{w})$ must be a Weyl–invariant function of the w and $\bar{w}$. This is because the interchange $z_I\leftrightarrow z_J$ induces (via (6.3.9b)) the following transformation in the w variables

$$w\rightarrow\sigma_r(w)\equiv w-(r\cdot w)\,r \tag{6.3.33}$$

where $\boldsymbol{r} = \boldsymbol{u}^{(I)} - \boldsymbol{u}^{(J)}$. Since the roots $\boldsymbol{r}$ are normalized with $(\boldsymbol{r})^2 = 2$, this transformation is simply the *Weyl reflection* of $\boldsymbol{w}$ in the hyperplane perpendicular to the root $\boldsymbol{r} = \boldsymbol{u}^{(I)} - \boldsymbol{u}^{(J)}$. To verify (6.3.33) one needs the properties (6.3.15) of the $\boldsymbol{u}^{(I)}$ vectors. (Note also that there are $\frac{1}{2}N(N-1)$ interchanges $z_I \leftrightarrow z_J$ and there are also $\frac{1}{2}N(N-1)$ Weyl reflections for $SU(N)$, one for each positive root.)

This implies that $G(\boldsymbol{w}, \bar{\boldsymbol{w}})$ must be invariant under all $\frac{1}{2}N(N-1)$ Weyl reflections. For the *Ansatz* (6.3.27), the factor $g\left((eB/2)\sum_{i=1}^{N-1}|w_i|^2\right)$ is indeed Weyl-invariant because $\sum_{i=1}^{N-1}|w_i|^2$ is itself Weyl-invariant. To see this, recall that the weights in an irreducible representation of $SU(N)$ may be generated by repeatedly reflecting the highest (dominant) weight Λ in all the positive roots. Therefore,

$$\sum_{\mu\in\{\Lambda\}} |\mu\cdot\boldsymbol{w}|^2 \quad , \tag{6.3.34}$$

(where the sum is over all weights μ in the representation $\{\Lambda\}$) is Weyl-invariant since under the Weyl reflection (6.3.33)

$$\begin{aligned}\mu\cdot\boldsymbol{w} &\to \mu\cdot(\boldsymbol{w} - (\boldsymbol{r}\cdot\boldsymbol{w})\,\boldsymbol{r}) \\ &= (\mu - (\boldsymbol{r}\cdot\mu)\,\boldsymbol{r})\cdot\boldsymbol{w} \\ &= \sigma_{\boldsymbol{r}}(\mu)\cdot\boldsymbol{w} \quad .\end{aligned}$$

In particular, for the adjoint representation the weights are just the roots themselves and so $\sum_{\boldsymbol{r}>0}|\boldsymbol{r}\cdot\boldsymbol{w}|^2$, and by (6.3.30) also $\sum_{i=1}^{N-1}|w_i|^2$, is Weyl-invariant.

The monomial factor

$$M_{\{\mathrm{d}\}}(\boldsymbol{w}) \equiv w_1^{d_1}w_2^{d_2}\ldots w_{N-1}^{d_{N-1}} \tag{6.3.35}$$

in (6.3.27) is *not necessarily* Weyl-invariant and must be explicitly symmetrized in the z variables. From the above discussion, they may be achieved by considering

$$M^{(d)}(\boldsymbol{w}) = \sum_{\text{all }\boldsymbol{r}} M_{\{\mathrm{d}\}}\left(\sigma_{\boldsymbol{r}}(\boldsymbol{w})\right) \tag{6.3.36}$$

which is a homogeneous polynomial of degree $d \equiv \sum_{i=1}^{N-1} d_i$ in the w_i variables. Note that in (6.3.36) the summation is over *all* roots of $SU(N)$; however, $M^{(d)}$ vanishes if d_1 is odd and we need only sum over the positive roots when d_1 is even.

Finally, collecting together the results (6.3.22), (6.3.25), (6.3.27), (6.3.32) and (6.3.35), and restoring the exponential factor, we arrive at a new family of Type-I eigenstates

$$\begin{aligned}\psi_{\mathrm{I}} &= \prod_{I<J}^{N}(z_{IJ})^{\nu}Z^{a}L_{\lambda_1}^{a}\left(\frac{eB}{2}|Z|^2\right) \\ &\times M^{(d)}(\boldsymbol{w})\;L_{\lambda_2}^{d+N-2+\frac{\nu}{2}N(N-1)}\left(\frac{eB}{2}\sum_{i=1}^{N-1}|w_i|^2\right)\exp\left(-\frac{eB}{4}\sum_{I=1}^{N}|z_I|^2\right) \quad .\end{aligned}$$

with

$$
\begin{aligned}
E_{\mathrm{I}} &= \omega\left(\frac{N}{2}+\lambda_1+\lambda_2\right) \\
J_{\mathrm{I}} &= \frac{\nu}{2}N(N-1)+a+d \ .
\end{aligned}
\tag{6.3.37a}
$$

An analogous derivation for Type-II states yields

$$
\psi_{\mathrm{II}} = \prod_{I<J}^{N}(\bar{z}_{IJ})^{2-\nu}\bar{Z}^{a}L^{a}_{\lambda_1}\left(\frac{eB}{2}|Z|^2\right)
\times M^{(d)}(\bar{w})\cdot L_{\lambda_2}^{d+N-2+\frac{2-\nu}{2}N(N-1)}\left(\frac{eB}{2}\sum_{i=1}^{N-1}|w_i|^2\right)\exp\left(-\frac{eB}{4}\sum_{I=1}^{N}|z_I|^2\right)
$$

with

$$
\begin{aligned}
E_{\mathrm{II}} &= \omega\left(\frac{N}{2}+\lambda_1+\lambda_2+a+d+\frac{2-\nu}{2}N(N-1)\right) \\
J_{\mathrm{II}} &= -\frac{(2-\nu)}{2}N(N-1)-a-d \ .
\end{aligned}
\tag{6.3.37b}
$$

These results (6.3.37) generalize to the N-anyon case the results of Y.-S. Wu (Wu 1984b) for the three-anyon problem in a harmonic potential. Moreover, if the parameter a in (6.3.37) is negative, we obtain the "mixed" states introduced in (Dunne *et al.* 1991a,b).

Finally, we note how the wavefunctions in (6.3.37a,b) are related to those in (6.3.6a,b). To make this comparison we use the following property of the generalized Laguerre polynomials (see Appendix A)

$$
L_n^{I_1+I_2+1}(x+y) = \sum_{\{\lambda_1+\lambda_2=n\}} L^{I_1}_{\lambda_1}(x)\cdot L^{I_2}_{\lambda_2}(y) \ , \tag{6.3.38}
$$

and observe that

$$
|Z|^2+\sum_{i=1}^{N-1}|w_i|^2 = \sum_{I=1}^{N}|z_I|^2 \ . \tag{6.3.39}
$$

Then, if one takes the wavefunction in (6.3.37a) and sums over all λ_1, λ_2 such that $\lambda_1+\lambda_2=n$ we find

$$
\begin{aligned}
&\prod_{I<J}^{N}(z_{IJ})^{\nu}Z^{a}M^{(d)}(w)\exp\left(-\frac{eB}{4}\sum_{I=1}^{N}|z_I|^2\right) \\
&\quad\times\sum_{\{\lambda_1+\lambda_2=n\}} L^{a}_{\lambda_1}\left(\frac{eB}{2}|Z|^2\right)L_{\lambda_2}^{d+N-2+\frac{\nu}{2}N(N-1)}\left(\frac{eB}{2}\sum_{i=1}^{N-1}|w_i|^2\right) \\
&=\prod_{I<J}^{N}(z_{IJ})^{\nu}Z^{a}M^{(d)}(w)\exp\left(-\frac{eB}{4}\sum_{I=1}^{N}|z_I|^2\right) \\
&\quad\times L_n^{a+d+\frac{\nu}{2}N(N-1)+N-1}\left(\frac{eB}{2}\sum_{I=1}^{N}|z_I|^2\right) \ ,
\end{aligned}
\tag{6.3.40}
$$

which is precisely the Type-I wavefunction in (6.3.6a) with the totally symmetric polynomial $\mathcal{S}^{(d_+)}$ separated into center-of-mass and relative coordinates. Thus the wavefunctions in (6.3.6) are special linear combinations of the more general wavefunctions presented in (6.3.37).

We conclude this section by observing that all the closed-form wavefunctions we have obtained by explicitly solving the differential equation (6.3.2) are special cases of the Type-I and Type-II states constructed in Section 2. For example all Type-I wavefunctions (6.3.37a) can be obtained by acting with suitable combinations of the step operators C_{0m}, C_{10} and C_{11} on the base state $\hat{\psi}_{\rm I}^{(0)}$ (Dunne *et al.* 1992a). We leave as an exercise for the reader to verify this assertion, but we want to point out that the step operator approach of Section 2 is much more general than the analytic one. Indeed all Type-I (6.2.17) involving C_{1m} with $m > 1$ do not appear within the closed-form solution we have presented in this section, but they are perfectly good and acceptable eigenstates of our Hamiltonian. Furthermore, the algebraic representation of the anyon wavefunctions in terms of step operators acting on base states will be extremely useful in computing the canonical partition functions and in analyzing the statistical mechanical properties of the anyon system, as we will see in Chapter 7. This analysis would be much more difficult using only the analytic methods related to the solutions of the differential equation (6.3.2).

6.4 Perturbation Theory and the Three-Anyon Problem

As we have already observed in Section 2, the set of Type-I and Type-II wavefunctions in (6.2.17) and (6.2.20) is not complete, and the bosonic and fermionic limits ($\nu \to 0, 1$) are not smooth due to missing states. To shed some light on the problem of these missing states, we adopt a perturbative approach: starting from an arbitrary bosonic or fermionic state, *all* of which are exactly known, we compute the perturbative corrections to its energy and wavefunction due to the non-trivial statistics (Chou 1991b; Mc Cabe and Ouvry 1991; Karlhede and Westerberg 1991; Khare and Mc Cabe 1991; Dasnières de Veigy and Ouvry 1991; Sporre *et al.* 1992b; Chou *et al.* 1992). In particular we show that the change in energy, ΔE, to first order in perturbation theory, obeys the following inequalities (Karlhede and Westerberg 1991)

$$-\tilde{\nu}\frac{N(N-1)}{2} \le \Delta E \le 0 \tag{6.4.1}$$

where

$$\tilde{\nu} = \begin{cases} \nu & \text{for perturbations around bosons} \\ \nu - 1 & \text{for perturbations around fermions} \end{cases} \tag{6.4.2}$$

and N is the number of particles. The Type-I states (6.2.17) are characterized by energy eigenvalues which are independent of ν (see (6.2.19a)), and thus for them

$$\Delta E_{\rm I} = 0 \ . \tag{6.4.3}$$

On the other hand, the Type-II states (6.2.20) have energies linearly dependent on ν (see (6.2.22a)), so that

$$\Delta E_{\rm II} = -\tilde{\nu}\frac{N(N-1)}{2} \quad . \tag{6.4.4}$$

As we will see in the following, $\Delta E = 0$ corresponds to the case when the relative angular momentum for *all* pairs of particles is non-negative, whereas $\Delta E = -\tilde{\nu}\, N(N-1)/2$ corresponds to the case when the angular momentum of *all* pairs is negative. Intermediate possibilities when the relative angular momenta for different pairs of particles do not have a definite sign, correspond to missing states whose wavefunctions are not of the form (6.1.18). The exact analytic expression of such wavefunctions is not known at the moment, but some approximate analytic solutions to this problem have recently become available (Chin and Hu 1992).

To prove (6.4.1) and apply perturbation theory, it is useful to write (see (6.3.1))

$$\hat{\psi} \equiv \prod_{I<J} (z_{IJ})^{\tilde{\nu}}\; P\,(z_1,\ldots,z_N;\bar{z}_1,\ldots\bar{z}_N) \quad , \tag{6.4.5}$$

and to consider the reduced Hamiltonian on the single-valued function P. The latter satisfies the differential equation (6.3.2) which we rewrite here for convenience

$$\sum_{I=1}^{N}\left(-\frac{2}{m}\partial_I + \omega\bar{z}_I\right)\bar{\partial}_I\, P - \frac{2}{m}\tilde{\nu}\sum_{I<J}\left(\frac{\bar{\partial}_I - \bar{\partial}_J}{z_I - z_J}\right) P = \left(E - \omega\frac{N}{2}\right)\, P \quad . \tag{6.4.6}$$

To further simplify the formulas, from now on we set $m = e\,B = 1$ and consider P to be symmetric or antisymmetric depending on whether we perturb around a bosonic or a fermionic state. As is clear from its definition (6.4.2), the parameter $\tilde{\nu}$ in (6.4.5) and (6.4.6) is the deviation from the bosonic or the fermionic statistics, and is taken to be the perturbation parameter. With these conventions (6.4.6) becomes

$$(\hat{H}_0 - \tilde{\nu}\,\mathcal{R})\, P = \left(E - \frac{N}{2}\right)\, P \tag{6.4.7}$$

where

$$\hat{H}_0 = \sum_{I=1}^{N} (-2\partial_I + \bar{z}_I)\,\bar{\partial}_I \tag{6.4.8a}$$

is the free Hamiltonian, and

$$\mathcal{R} = -2\sum_{I<J}\mathcal{R}_{IJ} = -2\sum_{I<J}\left(\frac{\bar{\partial}_I - \bar{\partial}_J}{z_I - z_J}\right) \tag{6.4.8b}$$

is the perturbation. For definiteness and simplicity, we will consider only the case in which we perturb a system of fermions, so that according to our conventions P must be completely antisymmetric. The perturbation theory around the bosonic point is more subtle and requires particular care (see for example (Chou *et al.* 1992)).

For $N = 1$ the eigenstates of the unperturbed Hamiltonian $\hat{H}_0$ are the Landau wavefunctions defined in (6.1.12) (up to the exponential factor that must be removed), namely

$$\hat{\psi}^j_n(z,\bar{z}) = \sqrt{\frac{n!}{\pi\, 2^{j+1}\,(n+j)!}}\; z^j\, L^j_n\left(\frac{1}{2}|z|^2\right) \quad . \tag{6.4.9}$$

They satisfy the orthonormality condition

$$\langle \hat{\psi}^{j_1}_{n_1} | \hat{\psi}^{j_2}_{n_2} \rangle \equiv \int d^2z\; \mathrm{e}^{-\frac{1}{2}|z|^2} \hat{\psi}^{*\,j_1}_{n_1}\; \hat{\psi}^{j_2}_{n_2} = \delta_{n_1 n_2}\, \delta_{j_1 j_2} \quad . \tag{6.4.10}$$

Note that the exponential factor which has been removed in the reduced wavefunctions, is restored inside the integral measure defining the inner product (6.4.10).

Since we consider perturbations around the fermionic point, the unperturbed N-body wavefunctions are *antisymmetrized* products of N Landau wavefunctions and can be written as Slater determinants, *i.e.*

$$\hat{\psi}_{n,j}(z_1,\ldots,z_N;\bar{z}_1,\ldots,\bar{z}_N) \propto \det \begin{pmatrix} \hat{\psi}^{j_1}_{n_1}(z_1) & \ldots & \hat{\psi}^{j_1}_{n_1}(z_N) \\ \vdots & \vdots & \vdots \\ \hat{\psi}^{j_N}_{n_N}(z_1) & \ldots & \hat{\psi}^{j_N}_{n_N}(z_N) \end{pmatrix} \tag{6.4.11}$$

where $n = \sum_{I=1}^{N} n_I$ and $j = \sum_{I=1}^{N} j_I$ are the total quantum numbers of the N-body state.

Since $\hat{\psi}_{n,j}$ is built out of a finite number of polynomials, it is a polynomial itself and furthermore it has a definite angular momentum j. Due to the high degeneracy of the problem, it is not immediately clear which fermionic polynomials admit a regular anyonic continuation; however, exploiting the conservation of angular momentum, we can diagonalize the perturbation $\mathcal{R}$ in the basis of the definite angular momentum functions $\hat{\psi}_{n,j}$.

To compute the expectation value of $\mathcal{R}$ for $\hat{\psi}_{n,j}$, and thus to find the first order corrections to the energy eigenvalues, it is sufficient to consider only one term of the sum defining $\mathcal{R}$, for example

$$\mathcal{R}_{12} = \frac{\bar{\partial}_1 - \bar{\partial}_2}{z_1 - z_2} \tag{6.4.12}$$

which depends only on the relative motion of particles 1 and 2. In fact, introducing $Z = (1/\sqrt{2})(z_1 + z_2)$ and the relative coordinate $z = z_1 - z_2$, we can rewrite $\mathcal{R}_{12}$ as follows

$$\mathcal{R}_{12} = 2\frac{\bar{\partial}_z}{z} \quad . \tag{6.4.13}$$

Since $\hat{\psi}_{n,j}$ is a polynomial in $z_1,\ldots,z_N$ and their complex conjugates, it is also a polynomial in $z, Z, z_3,\ldots,z_N$ and their complex conjugates; moreover the relative coordinate dependence can be conveniently factorized and we can write

$$\hat{\psi}_{n,j} = \sum_{n_0,j_0} \hat{\psi}^{j_0}_{n_0}(z,\bar{z})\ \hat{\phi}_{n-n_0,j-j_0}(Z,z_3,\dots,z_N;\bar{Z},\bar{z}_3,\dots,\bar{z}_N) \qquad (6.4.14)$$

where $\hat{\psi}^{j_0}_{n_0}$ is the normalized Landau wavefunction (6.1.12) depending only on the relative coordinate of particles 1 and 2, and $\hat{\phi}_{n-n_0,j-j_0}$ is some (not necessarily normalized) energy and angular momentum eigenfunction depending on the remaining degrees of freedom. If $\hat{\psi}_{n,j}$ is normalized, from (6.4.10) it follows that

$$\sum_{n_0,j_0} \langle \hat{\phi}_{n-n_0,j-j_0} | \hat{\phi}_{n-n_0,j-j_0} \rangle = 1 \ . \qquad (6.4.15)$$

Using the decomposition (6.4.14) and observing that the $\hat{\phi}$'s are orthonormal because they have different energies and angular momenta, we immediately find

$$\langle \hat{\psi}_{n,j} | \mathcal{R}_{12} | \hat{\psi}_{n,j} \rangle = \sum_{n_0,j_0} \langle \hat{\psi}^{j_0}_{n_0} | \mathcal{R}_{12} | \hat{\psi}^{j_0}_{n_0} \rangle \cdot \langle \hat{\phi}_{n-n_0,j-j_0} | \hat{\phi}_{n-n_0,j-j_0} \rangle \ . \qquad (6.4.16)$$

The matrix element $\langle \hat{\psi}^{j_0}_{n_0} | \mathcal{R}_{12} | \hat{\psi}^{j_0}_{n_0} \rangle$ can be easily computed using the properties of the generalized Laguerre polynomials reported in the Appendix. We leave the explicit calculation as an exercise to the reader and we report here only the extremely simple result

$$\langle \hat{\psi}^{j_0}_{n_0} | \mathcal{R}_{12} | \hat{\psi}^{j_0}_{n_0} \rangle = \begin{cases} 0 & \text{if } j_0 \geq 0 \\ \frac{1}{2} & \text{if } j_0 < 0 \end{cases} \qquad (6.4.17)$$

Inserting this into (6.4.16), we get

$$\langle \hat{\psi}_{n,j} | \mathcal{R}_{12} | \hat{\psi}_{n,j} \rangle = \frac{1}{2} \sum_{n_0} \sum_{j_0<0} \langle \hat{\phi}_{n-n_0,j-j_0} | \hat{\phi}_{n-n_0,j-j_0} \rangle \ . \qquad (6.4.18)$$

Due to the symmetries of the problem (the full perturbation $\mathcal{R}$ is symmetric under particle exchanges, while $\hat{\psi}^j_n$ is antisymmetric) the total first order energy correction ΔE is

$$\begin{aligned} \Delta E &= \langle \hat{\psi}_{n,j} | (-\tilde{\nu}\,\mathcal{R}) | \hat{\psi}_{n,j} \rangle = -\tilde{\nu}\, N(N-1) \langle \hat{\psi}_{n,j} | \mathcal{R}_{12} | \hat{\psi}_{n,j} \rangle \\ &= -\tilde{\nu}\, \frac{N(N-1)}{2} \sum_{n_0} \sum_{j_0<0} \langle \hat{\phi}_{n-n_0,j-j_0} | \hat{\phi}_{n-n_0,j-j_0} \rangle \ . \end{aligned} \qquad (6.4.19)$$

Each term in the sum appearing in the last line of (6.4.19) is positive definite, thus

$$\Delta E \leq 0 \ . \qquad (6.4.20a)$$

On the other hand the maximum value of this sum is 1 (see (6.4.15)), so that

$$\Delta E \geq -\tilde{\nu}\, \frac{N(N-1)}{2} \ . \qquad (6.4.20b)$$

This concludes the proof of (6.4.1).

The bound (6.4.20a) is saturated (*i.e.* $\Delta E = 0$) only if

$$\langle \hat{\phi}_{n-n_0,j-j_0} | \hat{\phi}_{n-n_0,j-j_0} \rangle = 0 \tag{6.4.21}$$

for all $j_0 < 0$. If this is true, the wavefunction $\hat{\psi}_{n,j}$ is decomposed with respect to *all* pairs of particles in relative wavefunctions with non-negative angular momenta. As anticipated at the beginning of this section, this occurs precisely for the Type-I wavefunctions (6.2.17) constructed with the step operator approach. The other bound (6.4.20b) is saturated (*i.e.* $\Delta E = -\tilde{\nu}\, N(N-1)/2$) when

$$\langle \hat{\phi}_{n-n_0,j-j_0} | \hat{\phi}_{n-n_0,j-j_0} \rangle = 1 \tag{6.4.22}$$

for all $j_0 < 0$. This means that $\hat{\psi}_{n,j}$ contains relative wavefunctions for *all* pairs of particles whose angular momentum is always negative. As previously remarked, this is precisely the case of the regular Type-II wavefunctions (6.2.20).

This analysis gives us a nice physical interpretation of the two families of anyonic wavefunctions which we have constructed using the step operators C_{nm}, and also furnishes some intuition on the nature of the missing states. The latter can indeed be characterized by saying that the relative angular momentum does not have a well defined sign for all pairs of particles. Also form this point of view, it is clear why the two-body problem (in which there is obviously only one pair) is completely solved by the Type-I and Type-II states of section 2 without missing states.

We now specialize our discussion to the three-body case and compute in particular the first order perturbative corrections to a fermionic wavefunction whose anyonic continuation is missing from our Type-I and Type-II states. To this aim it is convenient to use the Jacobi coordinates Z, w_1, w_2 defined in (6.3.8), so that the differential equation (6.4.6) becomes

$$\left(\hat{H}_{\text{c.m.}} + \hat{H}_{\text{rel}}\right) P = \left(E - \frac{3}{2}\right) P \tag{6.4.23}$$

where

$$\begin{aligned}
\hat{H}_{\text{c.m.}} &= \left(-2\partial_Z + \bar{Z}\right) \bar{\partial}_Z \quad , \\
\hat{H}_{\text{rel}} &= \sum_{i=1}^{2} \left(-2\partial_{w_i} + \bar{w}_i\right) \bar{\partial}_{w_i} \\
&\quad - 2\tilde{\nu} \left(\frac{\bar{\partial}_{w_1}}{w_1} + \frac{\bar{\partial}_{w_1} + \sqrt{3}\bar{\partial}_{w_2}}{w_1 + \sqrt{3} w_2} + \frac{\bar{\partial}_{w_1} - \sqrt{3}\bar{\partial}_{w_2}}{w_1 - \sqrt{3} w_2} \right) \quad .
\end{aligned} \tag{6.4.24}$$

The center of mass dynamics governed by $\hat{H}_{\text{c.m.}}$ is trivial and decouples from the relative motion. From now on we will neglect the center of mass contributions both to the wavefunctions and to the energy spectrum, and will focus only on the problem of the relative motion. As we did before, we treat the $\tilde{\nu}$-dependent terms of $\hat{H}_{\text{rel}}$ in (6.4.24) as a perturbation, and write

$$\hat{H}_{\text{rel}} = \hat{H}^0_{\text{rel}} - 2\tilde{\nu}\, \mathcal{R}(w_1, w_2; \bar{w}_1, \bar{w}_2) \quad . \tag{6.4.25}$$

We are interested in particular in the perturbative corrections to the non degenerate fermionic state described by

$$P^0 = \frac{1}{2}(w_1^* w_2 - w_1 w_2^*) \ . \tag{6.4.26}$$

Despite its appearance this wavefunction is completely antisymmetric under all particle exchanges. In terms of the Jacobi coordinates the exchange of particles 1 and 2 corresponds to

$$\begin{aligned} w_1 &\longrightarrow -w_1 \ , \\ w_2 &\longrightarrow w_2 \ ; \end{aligned} \tag{6.4.27a}$$

the exchange of particles 1 and 3 corresponds to

$$\begin{aligned} w_1 &\longrightarrow \frac{w_1 - \sqrt{3}w_2}{2} \ , \\ w_2 &\longrightarrow \frac{-\sqrt{3}w_1 - w_2}{2} \ ; \end{aligned} \tag{6.4.27b}$$

and finally the exchange of particles 2 and 3 corresponds to

$$\begin{aligned} w_1 &\longrightarrow \frac{w_1 + \sqrt{3}w_2}{2} \ , \\ w_2 &\longrightarrow \frac{\sqrt{3}w_1 - w_2}{2} \ . \end{aligned} \tag{6.4.27c}$$

It is immediate to verify that (6.4.26) is indeed antisymmetric under all transformations (6.4.27). Up to a numerical factor, the wavefunction (6.4.26) is the same as that in (6.1.26) written in Jacobi coordinates, and in the closely related problem of three anyons in a harmonic oscillator potential it represents the non-degenerate fermionic ground state (Wu 1984b).

Writing

$$\begin{aligned} P &= P^0 + \tilde{\nu} P^1 + \tilde{\nu}^2 P^2 + \dots \ , \\ E &= E^0 + \tilde{\nu} E^1 + \tilde{\nu}^2 E^2 + \dots \ , \end{aligned} \tag{6.4.28}$$

and equating the coefficients of the different powers of $\tilde{\nu}$ in the eigenstate equation for the relative motion [21]

$$\left[\hat{H}^0_{\rm rel} - 2\tilde{\nu}\,\mathcal{R}(w_1, w_2; \bar{w}_1, \bar{w}_2)\right] P = (E-1)\,P \ ,$$

we get

$$(\hat{H}^0_{\rm rel} - E^0 + 1)\,P^0 = 0 \ , \tag{6.4.29a}$$

$$(\hat{H}^0_{\rm rel} - E^0 + 1)\,P^1 - (2\mathcal{R}(w_1, w_2; \bar{w}_1, \bar{w}_2) + E^1)\,P^0 = 0 \ , \tag{6.4.29b}$$

$$(\hat{H}^0_{\rm rel} - E^0 + 1)\,P^2 - (2\mathcal{R}(w_1, w_2; \bar{w}_1, \bar{w}_2) + E^1)\,P^1 - E^2\,P^0 = 0 \tag{6.4.29c}$$

$$\vdots$$

[21] If also the the center of mass is considered, $(E-1)$ must be replaced by $(E-3/2)$ in the right hand side.

Using the explicit expressions of $\hat{H}^0_{\rm rel}$ and P^0, we can easily solve (6.4.29a) and find

$$E^0 = 2 \quad .$$

The second equation allows to determine E^1 and also P^1, but only up to terms proportional to P^0 because of (6.4.29a). We leave as an exercise to the reader to verify that

$$\begin{aligned} P^1 =& \frac{1}{2}\Big\{\bar{w}_1 w_2 \ln(w_1\bar{w}_1) \\ & - (\sqrt{3}\bar{w}_2 - \bar{w}_1)(\sqrt{3}w_1 + w_2)\ln[(\sqrt{3}w_1 + w_2)(\sqrt{3}\bar{w}_1 + \bar{w}_2)] \\ & - (\sqrt{3}\bar{w}_2 + \bar{w}_1)(\sqrt{3}w_1 + w_2)\ln[(\sqrt{3}w_2 - w_1)(\sqrt{3}\bar{w}_2 - \bar{w}_1)]\Big\} \quad , \\ E^1 =& -\frac{3}{2} \end{aligned} \tag{6.4.30}$$

solve (6.4.29b). Thus, to first order we have (Karlhede and Westerberg 1991)

$$\begin{aligned} E &= 2 - \frac{3}{2}\tilde{\nu} \quad , \\ P &= P^0 + \tilde{\nu} P^1 \quad . \end{aligned} \tag{6.4.31}$$

Unfortunately this solution cannot be extended in analytic form beyond the first order. We note that the energy eigenvalue in (6.4.31) has a negative $\tilde{\nu}$-derivative; its value $(-3/2)$ corresponds to that of the Type-II wavefunctions of Section 2 (see (6.2.20) for $N = 3$). One is therefore tempted to identify (6.4.31) with the perturbative expansion of some Type-II states. However this cannot be the case because, as we have already remarked, the state described by P^0 is not contained in the fermionic limit of (6.2.20). This means that the anyonic continuation of P^0 has higher perturbative corrections and its energy depends *non-linearly* on $\tilde{\nu}$. This expectation is indeed confirmed by recent numerical investigations (Sporre *et al.* 1991a; Murphy *et al.* 1991; Sporre *et al.* 1991b).

6.A Appendix

In this appendix we list some of the main properties of the generalized Laguerre polynomials which are useful to derive the results presented in the previous sections. The generalized Laguerre polynomial, $L^j_n(x)$, satisfies the differential equation

$$x\, f'' + (j + 1 - x)\; f' + n\, f = 0 \quad . \tag{6.A.1}$$

The Rodriguez formula for L^j_n is

$$L^j_n(x) = \frac{1}{n!} e^x x^{-j} \frac{d^n}{dx^n}\left(e^{-x} x^{n+j}\right) \quad , \tag{6.A.2}$$

and L^j_n is a polynomial for $j \geq -n$, $n \in \mathbf{Z}^+$. Some special cases are

$$L_0^j = 1 \ ,$$
$$L_1^j = j + 1 - x \ ,$$
$$L_n^0 = L_n \ , \quad \text{the ordinary Laguerre polynomial} \ ,$$
$$L_n^{-n} = \frac{(-1)^n}{n!} x^n \ .$$

They form an orthogonal set for each $j > -1$,

$$\int_0^\infty dx\, e^{-x} x^j L_n^j(x) L_m^j(x) = \frac{(n+j)!}{n!} \delta_{nm} \ . \tag{6.A.3}$$

An explicit expansion for L_n^j is

$$L_n^j(x) = \sum_{k=0}^{n} \binom{n+j}{n-k} \frac{(-x)^k}{k!} \ . \tag{6.A.4}$$

The L_n^j satisfy a two-term recurrence relation

$$n\, L_n^j(x) = (2n + j - 1 - x)\, L_{n-1}^j(x) - (n + j - 1)\, L_{n-2}^j(x) \tag{6.A.5}$$

for $n = 2, 3, 4, \ldots$. When the upper index is a negative integer, we have

$$L_n^{-j}(x) = (-x)^j \frac{(n-j)!}{n!} L_{n-j}^j(x) \ . \tag{6.A.6}$$

The polynomials L_n^j satisfy the following differential-difference relations

$$\frac{d}{dx} L_n^j = -L_{n-1}^{j+1} \ , \tag{6.A.7a}$$
$$x \frac{d}{dx} L_n^j = n L_n^j - (n+j) L_{n-1}^j \ , \tag{6.A.7b}$$

Finally, the generalized Laguerre polynomials satisfy the following addition formula,

$$L_n^{\alpha_1 + \ldots + \alpha_k + k - 1}(x_1 + \ldots + x_k) = \sum_{\{i_1 + \ldots + i_k = n\}} L_{i_1}^{\alpha_1}(x_1) \ldots L_{i_k}^{\alpha_k}(x_k) \ . \tag{6.A.8}$$

7. Statistical Mechanics of Anyons

In this chapter we are going to study some statistical mechanical properties of systems of many anyons, and in particular we will discuss the partition functions, the virial coefficients, the magnetization and the magnetic moment. However only very partial results can be obtained in this context, because the exact solution of a gas of anyons is not known. In fact, in contrast to the bosonic or fermionic case where the statistics is implemented by hand on the many body Hilbert space by constructing completely symmetric or antisymmetric products of single particle wavefunctions, for anyons the complicated boundary conditions for the interchange of any two particles require the knowledge of the *complete* many-body configurations. As we have pointed out in the previous chapter, only the two-body problem is exactly soluble for anyons, and hence only the two-body partition function can be computed exactly. Since the thermodynamic limit cannot be performed, one has to resort to approximate or alternative methods to study the statistical mechanics of anyons. For example if the thermodynamic functions are analytic in the particle density, it is well-known that the low density, or equivalently the high temperature limit, of a (free) gas can be investigated using the virial expansion (see for instance (Huang 1987; Reichl 1980)) in which the equation of state that relates the pressure P, the temperature T and the density ρ, can be expressed as follows

$$P = \rho\, k_{\mathrm{B}}\, T\, \left(1 + a_2\, \rho + a_3\, \rho^2 + \ldots\right) \tag{7.1}$$

where k_{B} is the Boltzmann constant and $a_2, a_3, \ldots$ are the so-called virial coefficients. For bosons and fermions, (7.1) is known to be consistent and one can assume that a similar expansion holds also in the case of particles with fractional statistics.

When $a_2, a_3, \ldots$ are all vanishing, the equation of state is that of a classical ideal gas ($P = \rho\, k_{\mathrm{B}}\, T$). The second virial coefficient a_2 depends only on the two-body quantum correlations, and it is positive for bosons and negative for fermions, signaling the presence of a statistical attraction or repulsion respectively. Since the two-body problem is completely soluble for anyons, the second virial coefficient can be computed exactly also for intermediate statistics (Arovas *et al.* 1985), and it turns out that it continuously interpolates between the negative bosonic value and the positive fermionic one. The third virial coefficient a_3 depends on the quantum two- and three-body correlations. It is known for bosons and fermions, but not for anyons since the three-anyon problem is not solved. However, even in the case of intermediate statistics, it is possible to derive some qualitative features of a_3 using semiclassical arguments (Bhaduri *et al.* 1991) or, as we will see, using the Type-I and Type-II states we constructed in the previous chapter (Dunne *et al.* 1992b). For the higher virial coefficients instead, only very preliminary and perturbative results are available at the moment (Comtet *et al.* 1991).

Remarkably enough, the exact solution of the two body problem allows to extract a few interesting features of the anyon system, like the presence or absence of spontaneous magnetization, or the presence or absence of parity violating effects, or the relation between the canonical magnetic moment and the fractional orbital angular momentum that we introduced in Chapter 5. However it would be even more remarkable and interesting to extend these results to the thermodynamic limit and deduce *macroscopic* effects that, at least in principle, could be detected. Lacking for the time being such possibility, we will content ourselves to compute the canonical partition functions and derive some consequences using standard techniques of statistical mechanics.

7.1 Partition Functions

The canonical N-body partition function, Z_N, for a system of free particles is clearly divergent because each energy level is infinitely degenerate. Thus, in order to get a finite Z_N, one has to break the degeneracy of the energy spectrum. This can be done, for instance, by confining the particles in a finite volume or, as we will do, by introducing a confining interaction, like a harmonic potential with frequency ω, which serves as a regulator. Since we will be interested in the magnetic properties of the anyon gas, we introduce also an external magnetic field B as we did in the previous chapter. The coupling to a magnetic field alone is not enough to regularize the theory and remove the degeneracy of the energy levels. Indeed, even though the presence of the magnetic field organizes the spectrum in discrete Landau levels spaced by the cyclotron frequency $\omega_c = eB/m$, each level remains infinitely degenerate in an infinite volume, thus yielding divergent partition functions. Therefore hereinafter we will consider a system of N particles in the infinite plane interacting both with a harmonic potential and with a magnetic field; sometimes the latter will be switched off to simplify the situation and render the formulas more manageable.

For bosons and fermions it is well-known how to compute the N-body partition functions Z_N. First one solves the single-particle problem and then forms *all* N-body wavefunctions by constructing all Slater permanents (for bosons) or determinants (for fermions) with the single-particle wavefunctions. In this way the energy levels and their degeneracies follow immediately, and one can easily compute Z_N. For example, for particles of mass m interacting in the plane only with a harmonic potential of frequency ω, the single-particle wavefunctions are (cf (6.1.12))

$$\psi_n^j \sim z^j \, L_n^j(m\omega|z|^2)\, \mathrm{e}^{-\frac{1}{2}m\omega|z|^2} \tag{7.1.1}$$

where $n = 0, 1, 2, \ldots$, $j = -n, -n+1, \ldots$ and L_n^j is the generalized polynomial of degree n (see Appendix 6.A). The energy corresponding to (7.1.1) is

$$E_n^j = \omega\ (2n + j + 1)\quad , \tag{7.1.2}$$

while the canonical one-body partition function is

$$Z_1 \equiv \sum_{\substack{n=0\\ j=-n}}^{\infty} \mathrm{e}^{-\beta E_n^j} \tag{7.1.3}$$

where, as usual, $\beta = 1/(k_B T)$. If we define $j' = j + n$ so that $j' = 0, 1, \ldots$ and $E_n^{j'} = \omega(n + j' + 1)$, we can easily evaluate the summations in (7.1.3) and get

$$\begin{aligned} Z_1 &= \sum_{\substack{n=0\\ j'=0}} \mathrm{e}^{-\beta E_n^{j'}} = \sum_{\substack{n=0\\ j'=0}} \mathrm{e}^{-\beta\omega(n+j'+1)} \\ &= \frac{\mathrm{e}^{-\beta\omega}}{(1-\mathrm{e}^{-\beta\omega})^2} = \frac{1}{4\sinh^2\left(\frac{\beta\omega}{2}\right)} \end{aligned} \quad . \tag{7.1.4}$$

Obviously Z_1 is the same for bosons, for fermions and for anyons because no statistical effect is present in the one-body problem.

Let us now compute the two-body partition function Z_2. For simplicity we first consider the case of bosons. The two-body bosonic wavefunctions are totally symmetric products of the single-particle wavefunctions (7.1.1). If (n_1, j_1') and (n_2, j_2') are respectively the labels of the two component factors, the total energy is

$$E_{n_1,n_2}^{j_1',j_2'} = \omega\ (n_1 + j_1' + n_2 + j_2' + 2) \quad , \tag{7.1.5}$$

and the partition function is

$$Z_2^{\mathrm{bos}} \equiv \sum_{\{n_1,j_1';n_2,j_2'\}_s} \mathrm{e}^{-\beta E_{n_1,n_2}^{j_1',j_2'}} \tag{7.1.6}$$

where the symbol $\sum_{\{n_1,j_1';n_2,j_2'\}_s}$ denotes the sum over all symmetric states meaning that the state characterized by the ordered set $\{n_1, j_1'; n_2, j_2'\}$ must be identified with the one characterized by $\{n_2, j_2'; n_1, j_1'\}$. To calculate Z_2 it is convenient to make the following observation: If we sum independently over all labels n_1, j_1', n_2 and j_2', we clearly overcount the bosonic states because we do not implement the identifications due to symmetry; however if we add to this sum the contribution of the diagonal states for which $n_1 = n_2$ and $j_1' = j_2'$, we get exactly twice the wanted bosonic states, that is

$$\sum_{\substack{n_1\\ j_1'}} \sum_{\substack{n_2\\ j_2'}} + \sum_{\substack{n_1=n_2\\ j_1'=j_2'}} = 2 \sum_{\{n_1,j_1';n_2,j_2'\}_s} \quad . \tag{7.1.7}$$

The bosonic partition function, Z_2^{bos}, then follows immediately from (7.1.7) and (7.1.5), and we get

$$
\begin{aligned}
Z_2^{\mathrm{bos}} &= \frac{1}{2}\left[\sum_{\substack{n_1=0\\ j_1'=0}}\sum_{\substack{n_2=0\\ j_2'=0}} \mathrm{e}^{-\beta\omega(n_1+j_1'+n_2+j_2'+2)} + \sum_{\substack{n_1=0\\ j_1'=0}} \mathrm{e}^{-\beta\omega(2n_1+2j_1'+2)}\right] \\
&= \frac{1}{2}\left[\frac{\mathrm{e}^{-2\beta\omega}}{(1-\mathrm{e}^{-\beta\omega})^4} + \frac{\mathrm{e}^{-2\beta\omega}}{(1-\mathrm{e}^{-2\beta\omega})^2}\right] \\
&= \frac{\cosh(\beta\omega)}{8\sinh^2\left(\frac{\beta\omega}{2}\right)\sinh^2(\beta\omega)} \;.
\end{aligned}
\tag{7.1.8}
$$

The fermionic two-body partition function

$$
Z_2^{\mathrm{fer}} \equiv \sum_{\{n_1,j_1';n_2,j_2'\}_{\mathrm{a}}} \mathrm{e}^{-\beta E_{n_1,n_2}^{j_1',j_2'}} \;, \tag{7.1.9}
$$

where the symbol $\sum_{\{n_1,j_1';n_2,j_2'\}_{\mathrm{a}}}$ denotes the sum over all antisymmetric states, can be computed in an analogous way. A moment thought reveals that

$$
\sum_{\substack{n_1\\ j_1'}}\sum_{\substack{n_2\\ j_2'}} - \sum_{\substack{n_1=n_2\\ j_1'=j_2'}} = 2 \sum_{\{n_1,j_1';n_2,j_2'\}_{\mathrm{a}}} \;. \tag{7.1.10}
$$

The diagonal states for which two quantum numbers are the same, are clearly impossible according to the Pauli exclusion principle, and thus must be removed. From (7.1.10), (7.1.9) and (7.1.5), it is easy to get

$$
\begin{aligned}
Z_2^{\mathrm{fer}} &= \frac{1}{2}\left[\sum_{\substack{n_1=0\\ j_1'=0}}\sum_{\substack{n_2=0\\ j_2'=0}} \mathrm{e}^{-\beta\omega(n_1+j_1'+n_2+j_2'+2)} - \sum_{\substack{n_1=0\\ j_1'=0}} \mathrm{e}^{-\beta\omega(2n_1+2j_1'+2)}\right] \\
&= \frac{1}{2}\left[\frac{\mathrm{e}^{-2\beta\omega}}{(1-\mathrm{e}^{-\beta\omega})^4} - \frac{\mathrm{e}^{-2\beta\omega}}{(1-\mathrm{e}^{-2\beta\omega})^2}\right] \\
&= \frac{1}{8\sinh^2\left(\frac{\beta\omega}{2}\right)\sinh^2(\beta\omega)} \;.
\end{aligned}
\tag{7.1.11}
$$

Let us now turn to the three-body partition functions for bosons and fermions. The bosonic three-body wavefunctions are totally symmetric products of three single-particle wavefunctions (7.1.1), and hence can be labeled by (n_1, j_1'), (n_2, j_2'), (n_3, j_3'). By taking into account the symmetry imposed by Bose statistics, it is not difficult to prove that, in obvious notations,

$$
\sum_{\substack{n_1\\ j_1'}}\sum_{\substack{n_2\\ j_2'}}\sum_{\substack{n_3\\ j_3'}} + 3\sum_{\substack{n_1=n_2\\ j_1'=j_2'}}\sum_{\substack{n_3\\ j_3'}} + 2\sum_{\substack{n_1=n_2=n_3\\ j_1'=j_2'=j_3'}} = 6\sum_{\{n_1,j_1';n_2,j_2'\}_{\mathrm{s}}} \;. \tag{7.1.12}
$$

Therefore, since

$$E_{n_1,n_2,n_3}^{j_1',j_2',j_3'} = \omega\,(n_1 + j_1' + n_2 + j_2' + n_3 + j_3' + 3)\ , \tag{7.1.13}$$

the partition function $Z_3^{\rm bos}$ is given by

$$\begin{aligned} Z_3^{\rm bos} &= \frac{1}{6}\left[\sum_{\substack{n_1=0\\ j_1'=0}}^{\infty}\sum_{\substack{n_2=0\\ j_2'=0}}^{\infty}\sum_{\substack{n_3=0\\ j_3'=0}}^{\infty} {\rm e}^{-\beta\omega(n_1+j_1'+n_2+j_2'+n_3+j_3'+3)}\right.\\ &\qquad\left.+3\sum_{\substack{n_1=0\\ j_1'=0}}^{\infty}\sum_{\substack{n_3=0\\ j_3'=0}}^{\infty} {\rm e}^{-\beta\omega(2n_1+2j_1'+n_3+j_3'+3)} + 2\sum_{\substack{n_1=0\\ j_1'=0}}^{\infty} {\rm e}^{-\beta\omega(3n_1+3j_1'+3)}\right]\\ &= \frac{1}{6}\left[\frac{{\rm e}^{-3\beta\omega}}{(1-{\rm e}^{-\beta\omega})^6} + 3\frac{{\rm e}^{-3\beta\omega}}{(1-{\rm e}^{-\beta\omega})^2(1-{\rm e}^{-2\beta\omega})^2} + 2\frac{{\rm e}^{-3\beta\omega}}{(1-{\rm e}^{-3\beta\omega})^3}\right]\\ &= \frac{\cosh(3\beta\omega) + 2\cosh^2\left(\frac{\beta\omega}{2}\right)}{32\sinh^2\left(\frac{\beta\omega}{2}\right)\sinh^2(\beta\omega)\sinh^2\left(\frac{3\beta\omega}{2}\right)}\ . \end{aligned} \tag{7.1.14}$$

In the case of fermions we must exclude all states in which at least two quantum numbrs of the constituent factors are the same; simple combinatorial considerations lead to the following result

$$\sum_{\substack{n_1\\ j_1'}}\sum_{\substack{n_2\\ j_2'}}\sum_{\substack{n_3\\ j_3'}} - 3\sum_{\substack{n_1=n_2\\ j_1'=j_2'}}\sum_{\substack{n_3\\ j_3'}} + 2\sum_{\substack{n_1=n_2=n_3\\ j_1'=j_2'=j_3'}} = 6\sum_{\{n_1,j_1';n_2,j_2'\}_{\rm a}}\ . \tag{7.1.15}$$

Using (7.1.13), we get

$$\begin{aligned} Z_3^{\rm fer} &= \frac{1}{6}\left[\sum_{\substack{n_1=0\\ j_1'=0}}^{\infty}\sum_{\substack{n_2=0\\ j_2'=0}}^{\infty}\sum_{\substack{n_3=0\\ j_3'=0}}^{\infty} {\rm e}^{-\beta\omega(n_1+j_1'+n_2+j_2'+n_3+j_3'+3)}\right.\\ &\qquad\left.-3\sum_{\substack{n_1=0\\ j_1'=0}}^{\infty}\sum_{\substack{n_3=0\\ j_3'=0}}^{\infty} {\rm e}^{-\beta\omega(2n_1+2j_1'+n_3+j_3'+3)} + 2\sum_{\substack{n_1=0\\ j_1'=0}}^{\infty} {\rm e}^{-\beta\omega(3n_1+3j_1'+3)}\right]\\ &= \frac{1}{6}\left[\frac{{\rm e}^{-3\beta\omega}}{(1-{\rm e}^{-\beta\omega})^6} - 3\frac{{\rm e}^{-3\beta\omega}}{(1-{\rm e}^{-\beta\omega})^2(1-{\rm e}^{-2\beta\omega})^2} + 2\frac{{\rm e}^{-3\beta\omega}}{(1-{\rm e}^{-3\beta\omega})^3}\right]\\ &= \frac{1 + 2\cosh^2\left(\frac{\beta\omega}{2}\right)}{32\sinh^2\left(\frac{\beta\omega}{2}\right)\sinh^2(\beta\omega)\sinh^2\left(\frac{3\beta\omega}{2}\right)}\ . \end{aligned} \tag{7.1.16}$$

This method can be easily applied to compute the bosonic and fermionic partition functions for any number N of particles, but the resulting formulas are more and more cumbersome when N increases. However, such a procedure cannot be applied to anyons of intermediate statistics. In fact, in this case, the N-body wavefunctions cannot be constructed from products of single-particle wave functions, because one has to represent the braid group and not simply the permutation group. An alternative approach is therefore needed. This is provided by the step operators introduced in Chapter 6 (Dunne *et al.* 1992b). The idea is to work in the anyon gauge in which case the Hamiltonian is identical for bosonic, fermionic or anyonic systems, but the many-body wavefunctions have different symmetry properties for different statistics. The Hamiltonian for N particles interacting with an external uniform magnetic field B and a harmonic force of frequency ω, is (in complex coordinates)

$$H = \sum_{I=1}^{N} \left(-\frac{2}{m}\partial_I \bar{\partial}_I + \frac{e^2 B^2}{8m}|z_I|^2 + \frac{1}{2}m\omega^2 |z_I|^2 \right) - \frac{eB}{2m} \sum_{I=1}^{N} \left(z_I \partial_I - \bar{z}_I \bar{\partial}_I \right) \ , \tag{7.1.17}$$

which is a simple generalization of (6.1.9). We are interested in the eigenvalue problem $H\psi = E\psi$ for multi-valued functions ψ. For convenience let us define

$$\Omega^2 \equiv 4\omega^2 + \frac{e^2 B^2}{m^2} = 4\omega^2 + \omega_c^2 \ , \tag{7.1.18}$$

and

$$\hat{\psi} = \exp\left(\frac{1}{4} m\Omega \sum_{I=1}^{N} |z_I|^2 \right) \psi \ , \tag{7.1.19}$$

so that the "reduced" Hamiltonian acting on $\hat{\psi}$ is

$$\hat{H} = \frac{N\Omega}{2} + \sum_{I=1}^{N} \left(-\frac{2}{m}\partial_I \bar{\partial}_I + \frac{1}{2}\omega_- z_I \partial_I + \frac{1}{2}\omega_+ \bar{z}_I \bar{\partial}_I \right) \ , \tag{7.1.20}$$

where $\omega_\pm = \Omega \pm \omega_c$. In analogy with (6.2.3a) the Hamiltonian $\hat{H}$ can be written as follows

$$\hat{H} = \frac{N\Omega}{2} + \frac{1}{2} \sum_{I=1}^{N} \left(\omega_+ \ a_I^\dagger a_I + \omega_- \ b_I^\dagger b_I \right) \ , \tag{7.1.21}$$

where the operators

$$\begin{aligned} a_I^\dagger &= \bar{z}_I - \frac{2}{m\Omega}\partial_I \ , \qquad & a_I &= \bar{\partial}_I \\ b_I^\dagger &= z_I - \frac{2}{m\Omega}\bar{\partial}_I \ , \qquad & b_I &= \partial_I \end{aligned} \tag{7.1.22}$$

satisfy the Heisenberg algebra $\left[a_I, a_J^\dagger\right] = \left[b_I, b_J^\dagger\right] = \delta_{IJ}$, with all other commutators being zero. This problem can be analyzed in complete analogy with the one discussed in Section 6.2. The symmetric step operators

$$C_{nm} = \sum_{I=1}^{N} a_I^{\dagger n} b_I^{\dagger m} \quad , \qquad (n+m \leq N) \tag{7.1.23}$$

act on the on the base states $\hat{\psi}_{\rm I}^{(0)} = \prod_{I<J} (z_{IJ})^{\nu}$, and $\hat{\psi}_{\rm II}^{(0)} = \prod_{I<J} (\bar{z}_{IJ})^{(2-\nu)}$, which entirely encode the anyonic statistics (here we take $0 \leq \nu < 2$), and the following two families of regular anyonic wavefunctions can be constructed

$$\hat{\psi}_{\rm I} = \prod_{n=0}^{1} \prod_{m=0}^{N-n} (C_{nm})^{\lambda_{nm}} \hat{\psi}_{\rm I}^{(0)} \tag{7.1.24a}$$

for Type-I, and

$$\hat{\psi}_{\rm II} = \prod_{m=0}^{1} \prod_{n=0}^{N-m} (C_{nm})^{\lambda_{nm}} \hat{\psi}_{\rm II}^{(0)} \tag{7.1.24b}$$

for Type-II. Here λ_{nm} are non-negative integers. For $N=2$ the states in (7.1.24), *completely* exhaust the space of wavefunctions for anyons of arbitrary statistics.

The energy corresponding to (7.1.24a) and (7.1.24b) is respectively

$$\begin{aligned} E_{\rm I}\left(\{\lambda_{nm}\}\right) =& \frac{N}{2}\Omega + \frac{1}{4}N(N-1)\omega_{-}\nu \\ &+ \frac{1}{2}\sum_{m=0}^{1}\sum_{n=0}^{N-m} \lambda_{nm}\left(n\omega_{+} + m\omega_{-}\right) \quad , \end{aligned} \tag{7.1.25a}$$

and

$$\begin{aligned} E_{\rm II}\left(\{\lambda_{nm}\}\right) =& \frac{N}{2}\Omega + \frac{1}{4}N(N-1)\omega_{+}(2-\nu) \\ &+ \frac{1}{2}\sum_{m=0}^{1}\sum_{n=0}^{N-m} \lambda_{nm}\left(n\omega_{+} + m\omega_{-}\right) \quad . \end{aligned} \tag{7.1.25b}$$

With the anyonic states represented as in (7.1.24), it is immediate to compute the corresponding partition functions. In fact, the contribution of the Type-I states (7.1.25a) to the partition function is

$$\begin{aligned} Z_N^{(\rm I)} &= \sum_{\{\lambda_{nm}\}} {\rm e}^{-\beta E_{\rm I}(\{\lambda_{nm}\})} \\ &= {\rm e}^{-\frac{N}{2}\beta\Omega - \frac{\beta}{4}N(N-1)\,\omega_{-}\nu} \prod_{n=0}^{1}\prod_{m=0}^{N-n}\sum_{\lambda_{nm}=0}^{\infty} {\rm e}^{-\frac{\beta}{2}(n\omega_{+}+m\omega_{-})\lambda_{nm}} \quad . \end{aligned} \tag{7.1.26}$$

The summations over λ_{nm} can be easily done and, with trivial manipulations, one gets

$$Z_N^{(\rm I)} = {\rm e}^{-\frac{N}{2}\beta\Omega - \frac{\beta}{4}N(N-1)\,\omega_{-}\nu}\, \frac{1}{2^{2N}} \prod_{\ell=1}^{N} \frac{1}{\sinh\left(\frac{\beta}{4}\omega_{-}\ell\right)\, \sinh\left[\frac{\beta}{4}\left(\omega_{-} + (\ell-1)\omega_{+}\right)\right]} \quad . \tag{7.1.27a}$$

The corresponding contribution $Z_N^{(\mathrm{II})}$ of the Type-II states (7.1.24b) to the partition function can be computed in the same way or simply obtained from (7.1.27a) with the substitutions $\omega_+ \leftrightarrow \omega_-$, $\nu \leftrightarrow 2-\nu$, yielding

$$Z_N^{(\mathrm{II})} = \mathrm{e}^{-\frac{N}{2}\beta\Omega - \frac{\beta}{4}N(N-1)\,\omega_+(2-\nu)} \frac{1}{2^{2N}} \prod_{\ell=1}^{N} \frac{1}{\sinh\left(\frac{\beta}{4}\omega_+\ell\right) \sinh\left[\frac{\beta}{4}\left(\omega_+ + (\ell-1)\omega_-\right)\right]} \,. \tag{7.1.27b}$$

Hence the total partition function for our boson-based anyons is given by

$$\begin{aligned} Z_N(\nu,B) &\equiv Z_N^{(\mathrm{I})} + Z_N^{(\mathrm{II})} \\ &= \frac{1}{2^{2N}} \left\{ e^{\frac{\beta}{4}N(N-1)\,\omega_-(1-\nu)} \prod_{\ell=1}^{N} \frac{1}{\sinh\left(\frac{\beta}{4}\omega_-\ell\right) \sinh\left[\frac{\beta}{4}\left(\omega_- + (\ell-1)\omega_+\right)\right]} \right. \\ &\quad \left. + e^{-\frac{\beta}{4}N(N-1)\,\omega_+(1-\nu)} \prod_{\ell=1}^{N} \frac{1}{\sinh\left(\frac{\beta}{4}\omega_+\ell\right) \sinh\left[\frac{\beta}{4}\left(\omega_+ + (\ell-1)\omega_-\right)\right]} \right\} \,. \end{aligned} \tag{7.1.28}$$

Note that this partition function enjoys the property $Z_N(\nu,B) = Z_N(2-\nu,-B)$ for any N and $Z_2(0,B) = Z_2(2,B)$ for the special case of two anyons. Moreover, we observe that Z_1 is obviously independent of ν, and that Z_2 in (7.1.28) is the complete partition function for two anyons of arbitrary statistics, and smoothly tends to the bosonic and fermionic ones for $\nu \to 0,2$ and $\nu \to 1$, respectively. For $N \geq 3$, $Z_N(\nu,B)$ does not reduce to the known bosonic and fermionic partition functions for $\nu = 0$ and $\nu = 1$, since the states (7.1.24) do not cover the entire spectrum.

Let us exemplify this for the special case of $B = 0$. In this case $\omega_+ = \omega_- = 2\omega$ and (7.1.28) gives

$$\begin{aligned} Z_2(\nu,0) &= \frac{\cosh\left[(1-\nu)\beta\omega\right]}{8\,\sinh^2\left(\frac{\beta\omega}{2}\right)\sinh^2(\beta\omega)} \,, \\ Z_3(\nu,0) &= \frac{\cosh\left[3(1-\nu)\beta\omega\right]}{32\,\sinh^2\left(\frac{\beta\omega}{2}\right)\sinh^2(\beta\omega)\,\sinh^2\left(\frac{3\beta\omega}{2}\right)} \,. \end{aligned} \tag{7.1.29}$$

By comparing these with (7.1.8) and (7.1.11), we see that $Z_2(0,0) = Z_2^{\mathrm{bos}}$ and $Z_2(1,0) = Z_2^{\mathrm{fer}}$. These equalities do not hold for $Z_3(\nu,0)$. However, let us note that the difference with respect to the exact results (7.1.14) and (7.1.16) is the same for $\nu = 0,1$. Indeed,

$$Z_3^{\mathrm{bos}} - Z_3(0,0) = \frac{\cosh^2\left(\frac{\beta\omega}{2}\right)}{16\,\sinh^2\left(\frac{\beta\omega}{2}\right)\sinh^2(\beta\omega)\sinh^2\left(\frac{3\beta\omega}{2}\right)} = Z_3^{\mathrm{fer}} - Z_3(1,0) \,. \tag{7.1.30}$$

Furthermore, we note that for $N = 3$, the coefficient of the contribution at first order in ν, *i.e.* $\frac{\partial}{\partial\nu} Z_3(\nu,0)\big|_{\nu=0}$, coincides exactly with the one computed with *numerical* methods in perturbation theory around bosonic statistics (McCabe and Ouvry 1991).

Having established that (7.1.28) is the complete partition function for two anyons in a magnetic field B and a confining harmonic potential of frequency ω, in the next sections we will compute the virial coefficients and the magnetization properties.

7.2 Virial Coefficients

The low density or equivalently the high temperature limit of a gas of particles can be described using the virial expansion, in which the equation of state is expressed as a series in the particle density ρ (see (7.1)). The coefficients in this expansion are known as virial coefficients. To set up the notations, let us first recall the basic definition of the first few of these. In the low density approximation, the gran canonical partition function $\mathcal{Q}$ can be written as a cluster expansion (see for instance (Huang 1987; Reichl 1980))

$$\mathcal{Q} = \exp\left(A\sum_{\ell=1}^{\infty} b_\ell\, z^\ell\right) \tag{7.2.1}$$

where b_ℓ are the so-called cluster integrals (Ursell 1927; Pais and Uhlenbeck 1959), z is the fugacity and A is the area of the system (we obviously limit our considerations to the case of two dimensions; in higher dimensions A would be replaced by the volume V). On the other hand, using standard techniques of statistical mechanics, the pressure P and the density ρ are given by

$$\begin{aligned} P &= k_{\mathrm{B}} T \left(\frac{1}{A}\ln\mathcal{Q}\right) , \\ \rho &= z\,\frac{\partial}{\partial z}\left(\frac{1}{A}\ln\mathcal{Q}\right)\bigg|_{A,T} . \end{aligned} \tag{7.2.2}$$

Upon inserting the cluster expansion (7.2.1) into (7.2.2), we get

$$\begin{aligned} P &= k_{\mathrm{B}} T \sum_{\ell=1}^{\infty} b_\ell\, z^\ell , \\ \rho &= \sum_{\ell=1}^{\infty} \ell\, b_\ell\, z^\ell , \end{aligned} \tag{7.2.3}$$

so that the virial expansion of the equation of state (7.1) becomes

$$\sum_{\ell=1}^{\infty} b_\ell\, z^\ell = \left(\sum_{\ell=1}^{\infty} \ell\, b_\ell\, z^\ell\right)\left[1 + a_2\left(\sum_{\ell=1}^{\infty} \ell\, b_\ell\, z^\ell\right) + a_3\left(\sum_{\ell=1}^{\infty} \ell\, b_\ell\, z^\ell\right)^2 + \ldots\right] \tag{7.2.4}$$

If we now expand both sides of (7.2.4) and equate the coefficients of equal powers of z, we easily obtain the expressions for the virial coefficients in terms of the cluster integrals; the first two are

$$a_2 = -\frac{b_2}{b_1^2} \,, \tag{7.2.5a}$$

$$a_3 = 4a_2^2 - 2\frac{b_3}{b_1^3} \,. \tag{7.2.5b}$$

We can make further progress if we use the definition of the gran canonical partition function

$$\mathcal{Q} = \sum_{N=0}^{\infty} z^N Z_N \tag{7.2.6}$$

to express the cluster integrals b_ℓ in terms of the N-body canonical partition functions Z_N. From the comparison of (7.2.1) with (7.2.6), we obtain

$$b_1 = \frac{Z_1}{A} \,, \tag{7.2.7a}$$

$$b_2 = \frac{2\,Z_2 - Z_1^2}{2\,A} \,, \tag{7.2.7b}$$

$$b_3 = \frac{3\,Z_3 - 3\,Z_2\,Z_1 + Z_1^3}{3\,A} \,. \tag{7.2.7c}$$

Actually all these expressions are meaningful only in the infinite area limit, so that particular care must be used in their evaluation.

In two spatial dimensions, one has

$$b_1 = \frac{1}{\lambda_{\mathrm{T}}^2} \tag{7.2.8}$$

where λ_{T} is the thermal wavelength of a particle of mass m and is given by $\lambda_{\mathrm{T}} = (2\pi\beta/m)^{1/2}$. Therefore from (7.2.7a) we have

$$\frac{1}{\lambda_{\mathrm{T}}^2} = \frac{Z_1}{A} \,. \tag{7.2.9}$$

Using the expression of Z_1 given in (7.1.4) and taking the high temperature limit (*i.e.* $\beta \to 0$), we are lead to the identification

$$\omega^2 \sim \frac{2\pi}{\beta m}\frac{1}{A} \,. \tag{7.2.10}$$

Thus the harmonic frequency ω can be thought as being related to the inverse of the area A, and hence the infinite area limit ($A \to 0$) becomes equivalent to the zero harmonic frequency limit ($\omega \to 0$). This is of course to be expected because putting the system in a harmonic potential confines the particles to move in a finite region; removing the harmonic force (*i.e.* letting $\omega \to 0$) amounts to remove any restriction on the space (*i.e.* $A \to 0$).

Let us now evaluate the cluster integrals and the first virial coefficients in the $\omega \to 0$ limit. For simplicity we will use the anyon partition functions Z_N in (7.1.28) with $B = 0$; the case with a non-vanishing magnetic field is left as an exercise for the reader.

Recalling that the harmonic regularization requires to multiply b_ℓ by a normalization factor $\ell^{d/2}$ where d is the number of space dimensions [22], the second virial coefficient turns out to be

$$\begin{aligned} a_2(\nu) &= -\frac{b_2}{b_1^2} = \lambda_T^2 \lim_{\omega\to 0}\left[Z_1 - 2\frac{Z_2(\nu,0)}{Z_1}\right] \\ &= \lambda_T^2\left(-\frac{1}{4} + \nu - \frac{1}{2}\nu^2\right) , \end{aligned} \tag{7.2.11}$$

in agreement with the original derivation (Arovas *et al.* 1985) which is based on different methods (cf also (Comtet *et al.* 1989; Blum *et al.* 1990; Sen 1991a)). We remark that both Z_1 and $2\,Z_2(\nu,0)/Z_1$ are divergent in the limit $\omega \to 0$ – in fact they are $O(1/\omega^2)$ – but the divergence cancels in the combination

$$Z_1 - 2\frac{Z_2(\nu,0)}{Z_1}$$

appearing in (7.2.11). The second virial coefficient $a_2(\nu)$ interpolates continuously between the bosonic value

$$a_2^{\rm bos} \equiv a_2(0) = -\frac{1}{4}\,\lambda_T^2 , \tag{7.2.12}$$

and the fermionic one

$$a_2^{\rm fer} \equiv a_2(1) = \frac{1}{4}\,\lambda_T^2 . \tag{7.2.13}$$

This is an obvious consequence of the fact that the two-body partition functions is exact for arbitrary statistics ν.

Let us now turn to the third virial coefficient which is given by

$$\begin{aligned} a_3(\nu) &= 4\,a_2(\nu)^2 - 2\,\frac{b_3}{b_1^3} \\ &= 4\,a_2(\nu)^2 - 2\,\lambda_T^4 \lim_{\omega\to 0}\left[3\,\frac{Z_3(\nu,0)}{Z_1} - 3\,Z_2(\nu,0) + Z_1^2\right] . \end{aligned} \tag{7.2.14}$$

[22]For two particles the harmonic potential is

$$\omega^2\,(z_1^2 + z_2^2) = 2\omega^2\,Z^2 + \frac{\omega^2}{2}\,(z_1^2 - z_2^2)$$

where $Z = (1/\sqrt{2})\,(z_1 + z_2)$ is the center of mass coordinate. Thus, in the two-body problem $\omega_{\rm c.m.}^2 = 2\,\omega^2$, and in general in the ℓ-body problem $\omega_{\rm c.m.}^2 = \ell\,\omega^2$. Since it is $\omega_{\rm c.m.}^2$ which is actually related to the area according to (7.2.10), the two-dimensional cluster integral b_ℓ must be multiplied by ℓ.

If we substitute in (7.2.14) the explicit expressions of the partition functions given in (7.1.29) and compute the $\omega \to 0$ limit, we find a divergent result. However, remarkably enough, the coefficients of the divergences turn out to be *independent of* ν, and thus the difference $a_3(\nu) - a_3(\nu')$ is well-defined and finite. Furthermore, a simple calculation shows that

$$a_3(\nu) - a_3(\nu') = 0 \ . \tag{7.2.15}$$

This result is the consequence of highly non-trivial cancellations and was first pointed out in (Dunne *et al.* 1992b).

We remark that the third virial coefficient is finite for bosons and fermions; indeed one has [23]

$$a_3^{\text{bos}} = a_3^{\text{fer}} = \frac{1}{36}\lambda_T^4 \ . \tag{7.2.16}$$

This result can be easily obtained if we use the *complete* bosonic or fermionic partition functions given in (7.1.14) and (7.1.16) respectively. As previously noted, these are not the smooth limit of $Z_3(\nu, 0)$ for $\nu \to 0, 1$, and in fact the difference given in (7.1.30) precisely cancels the divergences in the cluster integrals and renders the third virial coefficient finite.

The presence of divergences in a_3 for fractional statistics could signal either the failure of the virial expansion due to long-range statistical interactions, or the presence of "missing states" which are not included in (7.1.24). This last possibility seems to be suggested by recent numerical (Sporre *et al.* 1991a; Murphy *et al.* 1991) and perturbative (McCabe and Ouvry 1991; Comtet *et al.* 1991; Khare and McCabe 1991) calculations. To shed some light on this problem, it is therefore necessary to investigate in more detail the third virial coefficient $a_3(\nu)$. To this purpose, let us turn back to the three-anyon problem, described by the reduced Hamiltonian (see (7.1.20) for $B = 0$)

$$\hat{H} = 3\omega + \sum_{I=1}^{3} \left(-\frac{2}{m}\, \partial_I \bar{\partial}_I + \omega\, z_I\, \partial_I + \omega\, \bar{z}_I\, \bar{\partial}_I \right) \ . \tag{7.2.17}$$

As we have seen in Chapter 6, $\hat{H}$ can be nicely split into a center of mass part and a relative part by introducing the Jacobi coordinates (cf (6.3.8))

$$\begin{aligned} Z &= \frac{1}{\sqrt{3}}\,(z_1 + z_2 + z_3) \ , \\ w_1 &= \frac{1}{\sqrt{2}}\,(z_1 - z_2) \ , \\ w_2 &= \frac{1}{\sqrt{6}}\,(z_1 + z_2 - 2z_3) \ . \end{aligned} \tag{7.2.18}$$

Indeed we have

$$\hat{H} = \hat{H}_{\text{c.m.}} + \hat{H}_{\text{rel}}$$

[23] It is well-known that all odd virial coefficients, a_{2k+1}, are equal for bosons and fermions (Huang 1987).

where

$$\hat{H}_{\mathrm{c.m.}} = \omega + \left(-\frac{2}{m} \partial_Z \bar{\partial}_Z + \omega\, Z\, \partial_Z + \omega\, \bar{Z}\, \bar{\partial}_Z \right) \quad ,$$

$$\hat{H}_{\mathrm{rel}} = 2\omega + \sum_{i=1}^{2} \left(-\frac{2}{m} \partial_{w_i} \bar{\partial}_{w_i} + \omega\, w_i\, \partial_{w_i} + \omega\, \bar{w}_i\, \bar{\partial}_{w_i} \right) \quad . \tag{7.2.19}$$

From now on, we focus only on the relative problem and neglect the trivial center of mass dynamics. In analogy with (7.1.22) we introduce the operators

$$A_i^\dagger = \bar{w}_i - \frac{1}{m\omega} \partial_{w_i} \quad , \qquad A_i = \bar{\partial}_{w_i}$$

$$B_i^\dagger = w_i - \frac{1}{m\omega} \bar{\partial}_{w_i} \quad , \qquad B_i = \partial_{w_i} \tag{7.2.20}$$

for $i = 1, 2$, so that $\hat{H}_{\mathrm{rel}}$ becomes

$$\hat{H}_{\mathrm{rel}} = \omega \left[2 + \sum_{i=1}^{2} \left(A_i^\dagger\, A_i + B_i^\dagger\, B_i \right) \right] \quad . \tag{7.2.21}$$

The operators (7.2.20) satisfy the Heisenberg algebra

$$[A_i, A_j^\dagger] = [B_i, B_j^\dagger] = \delta_{ij} \quad , \tag{7.2.22}$$

with all other commutators being zero, and are energy step operators. In fact, using (7.2.21) and (7.2.22), it is easy to prove that A_i and B_i lower the eigenvalues of $\hat{H}_{\mathrm{rel}}$ by ω, *i.e.*

$$[\hat{H}_{\mathrm{rel}}, A_i] = -\omega\, A_i \quad , \qquad [\hat{H}_{\mathrm{rel}}, B_i] = -\omega\, B_i \quad , \tag{7.2.23a}$$

whereas $A_i^\dagger$ and $B_i^\dagger$ raise the eigenvalues by ω, *i.e.*

$$[\hat{H}_{\mathrm{rel}}, A_i^\dagger] = \omega\, A_i^\dagger \quad , \qquad [\hat{H}_{\mathrm{rel}}, B_i^\dagger] = \omega\, B_i^\dagger \quad . \tag{7.2.23b}$$

Following (Sen 1991b,c), we define the operator

$$Q = A_1^\dagger\, B_2 - A_2^\dagger\, B_1 \tag{7.2.24}$$

which is completely antisymmetric under particle exchanges, and commutes with $\hat{H}_{\mathrm{rel}}$. Being antisymmetric, Q acts on symmetric wavefunctions to produce antisymmetric ones with the same energy, and vice versa. Thus Q changes the statistics mapping bosons into fermions or vice versa, and for this reason it has been called a supersymmetry operator (Sen 1991b,c). More generally, Q connects staes of equal energy in two anyonic theories with statistics ν and $\nu+1$. In order to keep the statistics within the range $[0,1]$, we can follow the action of Q with a parity transformation P which makes the replacements $w_i \leftrightarrow \bar{w}_i$ in the wavefunctions, corresponding to $(x_1, y_i) \to (x_i, -y_i)$ in terms of the cartesian coordinates. Since P changes the sign of the statistical parameter, the combined operator

$$\tilde{Q} = PQ \tag{7.2.25}$$

maps anyonic states of statistics ν to anyonic states of statistics $1-\nu$ without changing their energy.

Let us now consider a wavefunction $\hat{\psi}(\nu)$ describing the relative motion of three anyons of statistics ν with a given energy $E(\hat{\psi})$. Under the action of $\tilde{Q}$, in principle there are three possibilities:

i) The wavefunction $\tilde{Q}\,\hat{\psi}(\nu)$ is a regular wavefunction for three anyons of statistics $1-\nu$ and energy $E(\hat{\psi})$;

ii) The wavefunction $\tilde{Q}\,\hat{\psi}(\nu)$ identically vanishes;

iii) The wavefunction $\tilde{Q}\,\hat{\psi}(\nu)$ is singular.

A detailed analysis is therefore necessary to see which one of these possibilities actually occurs. We refer the reader to the original literature (Sen 1991b,c) for a complete discussion; here we limit ourselves to summarize the results and see their consequences on the third virial coefficient. It turns out that if $\hat{\psi}(\nu)$ is a regular wavefunction, then case *iii)* never occurs, that is $\tilde{Q}\,\hat{\psi}(\nu)$ is always an acceptable wavefunction (eventually vanishing) of a new anyonic theory with statistics $1-\nu$. Furthermore, the operator $\tilde{Q}$ annihilates *only* all the Type-I and Type-II states which we constructed in Chapter 6 and which contributed to the partition function $Z_3(\nu,0)$ in (7.1.29). The missing states (if they exist) are instead mirror symmetric, that is for any regular missing wavefunction $\hat{\psi}'(\nu)$ there is a corresponding regular missing wavefunction $\hat{\psi}'(1-\nu)$. This property is confirmed also by recent numerical calculations (Sporre *et al.* 1991a; Murphy *et al.* 1991).

Let us now assume that the missing states exist and construct the *full* three-body partition function according to

$$\begin{aligned}\tilde{Z}_3(\nu) &\equiv \sum_{\{\text{all states } \hat{\psi}\}} e^{-\beta E(\hat{\psi})} \\ &= Z_3(\nu,0) + \sum_{\{\text{missing states } \hat{\psi}'\}} e^{-\beta E(\hat{\psi}')} \ .\end{aligned} \tag{7.2.26}$$

By construction we have

$$\tilde{Z}_3(\nu=0) = Z_3^{\text{bos}} \ , \tag{7.2.27a}$$

and

$$\tilde{Z}_3(\nu=1) = Z_3^{\text{fer}} \ , \tag{7.2.27b}$$

since the missing states are defined to make the bosonic and fermionic limits smooth. Taking into account the mirror symmetry of the missing states, we easily obtain

$$\tilde{Z}_3(\nu) - \tilde{Z}_3(1-\nu) = Z_3(\nu,0) - Z_3(1-\nu,0) \ , \tag{7.2.28}$$

which shows that only the regular Type-I and Type-II states contribute to the difference $\tilde{Z}_3(\nu) - \tilde{Z}_3(1-\nu)$. Let us now define

$$\tilde{a}_3(\nu) \equiv 4\,a_2(\nu)^2 - 2\,\lambda_T^4 \lim_{\omega\to 0}\left[3\,\frac{\tilde{Z}_3(\nu)}{Z_1} - 3\,Z_2(\nu,0) + Z_1^2\right] \quad , \tag{7.2.29}$$

which is the same expression as in (7.2.14) with $Z_3(\nu,0)$ replaced by the *full* partition function $\tilde{Z}_3(\nu)$, and then compute

$$\tilde{a}_3(\nu) - \tilde{a}_3(1-\nu) \ .$$

Using (7.2.28), we have

$$\begin{aligned}
\tilde{a}_3(\nu) - \tilde{a}_3(1-\nu) &= 4\,a_2(\nu)^2 - 4\,a_2(1-\nu)^2 \\
&\quad - 2\,\lambda_T^4 \lim_{\omega\to 0}\left\{3\left[\frac{\tilde{Z}_3(\nu) - \tilde{Z}_3(1-\nu)}{Z_1}\right]\right. \\
&\qquad\qquad \left. - 3\left[Z_2(\nu,0) - Z_2(1-\nu,0)\right]\right\} \\
&= a_3(\nu) - a_3(1-\nu) \ .
\end{aligned} \tag{7.2.30}$$

From (7.2.15) with $\nu' = 1-\nu$, it immediately follows that

$$\tilde{a}_3(\nu) = \tilde{a}_3(1-\nu) \ . \tag{7.2.31}$$

The mirror symmetry of the third virial coefficient expressed by (7.2.31) is a rigorous result which was first pointed out in (Sen 1991c). In particular for $\nu = 0$, we have the well-known property

$$a_3^{\rm bos} \equiv \tilde{a}_3(0) = \tilde{a}_3(1) \equiv a_3^{\rm fer} = \frac{1}{36}\,\lambda_T^4 \ . \tag{7.2.32}$$

Unfortunately, the mirror symmetry (7.2.31) does not allow to deduce the value of $\tilde{a}_3(\nu)$ for any ν; however, using (7.2.31) together with the perturbative calculations of (McCabe and Ouvry 1991), it can be *conjectured* that actualy the third virial coefficient does not depend on the statistics ν. A proof (or a disproof) of this conjecture is still missing.

For the higher virial coefficients, only partial pertubative results are available at the moment but no definite conclusions can be drawn (McCabe *et al.* 1991).

7.3 Magnetic Moment

In this section we are going to analyze some magnetic properties of the anyon gas. Let us consider a two-body system and define the magnetization according to the standard formula

$$M_2 = \frac{1}{\beta}\,\frac{\partial}{\partial B}\log Z_2 \ . \tag{7.3.1}$$

As a consequence of the properties of Z_2 given after (7.1.28), it is easy to see that M_2 satisfies

$$M_2(\nu, B) = -M_2(2-\nu, -B) \ . \tag{7.3.2}$$

There are two regimes in which one can extract simple and interesting results. The first is the regime of *high magnetic fields*, in which B sets the biggest scale in the problem, namely

$$\frac{m}{e\beta B} \ll 1 \ , \qquad \frac{\omega m}{eB} \ll 1 \ . \tag{7.3.3}$$

The first inequality tells us that the thermal excitations whose energies are $\sim 1/\beta$ are negligible with respect to the magnetic excitations whose energies are $\sim \omega_c = eB/m$. Instead the second equality in (7.3.3) implies that the excitations induced by the harmonic force whose energies are $\sim \omega$ are small compared to the magnetic excitations. In this regime M_2 reduces to a double series in the small parameters (7.3.3), whose first terms are given by

$$M_2(B,\nu) = \begin{cases} -2\frac{e}{2m}\left[1 - \frac{2m}{e\beta B} - \left(\nu + \frac{1}{2} - 2\delta_{\nu,2}\right)\left(\frac{\omega m}{eB}\right)^2\right] \ , & B > 0 \\ 2\frac{e}{2m}\left[1 - \frac{2m}{e\beta |B|} + \left(\nu - \frac{5}{2} + 2\delta_{\nu,0}\right)\left(\frac{\omega m}{eB}\right)^2\right] \ , & B < 0 \ . \end{cases} \tag{7.3.4}$$

For $\omega = 0$ the magnetic moment M_2 is independent of statistics; the second term on the right-hand side of (7.3.4) represents a finite temperature correction. As expected, the statistical dependence appears in the term proportional to ω; it is in fact the harmonic potential that breaks the infinite degeneracy of the Landau levels of two anyons. The last term in (7.3.4) can also be reinterpreted as a "finite area correction" if we take into account the identification of ω^2 with the inverse of the area as shown in (7.2.10). With such identification the last term of (7.3.4) represents a correction inversely proportional to the total flux BA through the sample.

We remark that M_2 in (7.3.4) satisfies

$$\begin{aligned} M_2\,(-B, \nu = 1) &= -M_2\,(B, \nu = 1) \\ M_2\,(-B, \nu = 0,2) &= -M_2\,(B, \nu = 0,2) \ . \end{aligned} \tag{7.3.5}$$

which show that *parity is not broken for bosons and fermions*; instead for all other values of ν parity is broken.

The second regime which leads to an interesting result is that of *low magnetic field*

$$\frac{\omega_c}{\omega} = \frac{eB}{m\omega} \ll 1 \ . \tag{7.3.6}$$

This is the regime in which one can take the limit $B \to 0$ for finite values of ω and β, and is the relevant one to study spontaneous magnetization effects. Indeed for low magnetic field, M_2 is given by a series in the small parameter (7.3.69) with functions of $\beta\omega$ as coefficients; more precisely we have

$$M_2(B,\nu) = \frac{e}{2m}\Big\{\mathrm{th}[\beta\omega(1-\nu)]\coth(\beta\omega) - (1-\nu)\Big\} + O\left(\frac{eB}{m\omega}\right) \ . \tag{7.3.7}$$

For intermediate statistics (*i.e.* for $\nu \neq 0,1,2$) there is a parity breaking *spontaneous magnetic moment* μ given by

$$\mu \equiv M_2(B=0,\nu) \ . \tag{7.3.8}$$

This spontaneous magnetic moment vanishes in the limit of high temperature,

$$\lim_{\beta \to 0} \mu = 0 \ , \tag{7.3.9}$$

whereas at zero temperature it reduces to

$$\mu_L \equiv \lim_{\beta \to \infty} \mu = \frac{e}{2m} L_\nu \tag{7.3.10}$$

where L_ν is the orbital fractional angular momentum defined in Chapter 5 (see in particular (5.85) and (5.86) for $N = 2$). Notice that the gyromagnetic factor for μ_L is 1, as usual for an *orbital* magnetic moment. This is to be contrasted with the spin magnetic moment μ_S computed in (Kogan 1991; Stern 1991) for which the gyromagnetic factor is 2, as should be expected. Finally, we remark that the spontaneous magnetic moment μ vanishes for non-confined anyons (*i.e.* when $\omega = 0$) at any finite temperature. It would be interesting and important to extend these results to systems of many anyons, and see if the spontaneous magnetization we have evidentiated for two anyons, survives in the thermodynamic limit. Another interesting open problem is the study of the possible existence of a Curie-like critical temperature and a corresponding phase transition; however to make progress in this direction, it is necessary first to solve the many-anyon problem.

8. Anyons and the Fractional Quantum Hall Effect

The only known physical objects which can be described as anyons are the quasi-particle and quasi-hole excitations of planar systems of electrons exhibiting the fractional quantum Hall effect (QHE) (for a review see for instance (Prange and Girvin 1990)). Actually most of the great interest that anyonic theories have attracted in the past few years derives precisely from their relevance to a better understanding of the fractional QHE (Halperin 1984), in conjunction with several claims that anyons can provide also a non-standard explanation of the mechanism of high temperature superconductivity (Chen *et al.* 1989). It is far beyond the scope of these lecture notes to treat these issues in a systematic and adequate way, and therefore we refer the reader to the many good reviews already existing in the literature, for example (Wen and Zee 1989b; Arovas 1989; Lykken *et al.* 1991). However, for the sake of completeness we think necessary to spend some time on at least one of these physical applications in order to convey the idea that anyons are not just mathematical fantasies. Since recent experiments have cast some shadow on the relevance of fractional statistics to the observed high temperature superconductivity (Lyons *et al.* 1990; Kiefl *et al.* 1990; Spielman *et al.* 1990), here we will concentrate only on the application of anyons to the theory of the fractional QHE. Of course our presentation will be schematic and not at all exhaustive. Furthermore, we will concentrate more on the formal aspects than on the condensed matter issues.

Let us recall that the QHE is observed in two dimensional systems of electrons at very low temperatures (a few degrees Kelvin or even less) and in very strong magnetic fields (~ 10 Tesla or even more) orthogonal to the plane where the particles move. The electrons are usually trapped in a thin layer at the interface between two different semiconductors or between a semiconductor and an insulator. The low temperature and the strong magnetic field freeze the motion along the direction perpendicular to the layer, and thus the relevant dynamics takes place on a plane; moreover the effects due to the electron spin can be neglected in a first approximation. The QHE is characterized by the fact that the Hall conductance, σ_{H}, is quantized in units of e^2/h ($-e$ is the electron charge and h is the Planck constant)

$$\sigma_{\mathrm{H}} = v\,\frac{e^2}{h} \quad . \tag{8.1}$$

The quantum number v can be an integer (integer QHE) (von Klitzing *et al.* 1980) or a simple fraction (fractional QHE) (Tsui *et al.* 1982). Quantization rules are usually associated to symmetry considerations, like for example the quantization of spin which is related to the algebra of rotations. The quantization rule of the QHE seems instead to have a different origin, at least in the integer case: it is a topological quantization (Thouless *et al.* 1982; Avron *et al.* 1983; Niu *et al.* 1985; Kohmoto 1985), like the one of the magnetic monopole charge (Dirac 1931). However, contrary to the latter which has not been observed so far, the quantization of $\sigma_{\rm H}$ is a well-established fact. Actually, what is astonishing and fascinating is that (8.1) is experimentally observed with extremely high accuracy (the precision is roughly $10^{-7} \sim 10^{-8}$ for the integer QHE, and $10^{-4} \sim 10^{-5}$ for the fractional QHE).

The external magnetic field organizes the energy spectrum of the electrons into Landau levels (see for instance (Landau and Lifschitz 1977) and Chapter 6) and forces the particles to fill such levels from bottom up. A quantity which plays a central role in the QHE is the filling factor υ [24], defined as the number of electrons, N, per number of Landau levels available. Each Landau level is highly degenerate (see (6.1.13)) but for samples of *finite* area A, this degeneracy is finite. The number of available levels is in fact given by

$$\frac{A}{2\pi\ell_0^2} \tag{8.2}$$

where ℓ_0 is the magnetic length

$$\ell_0 = \sqrt{\frac{\hbar c}{eB}} \quad . \tag{8.3}$$

Therefore the filling factor is

$$\upsilon = \frac{N}{A/2\pi\ell_0^2} = 2\pi\ell_0^2\rho \tag{8.4a}$$

where $\rho = N/A$ is the electron density. Eq. (8.4a) can also be written in a different way using the explicit definition of the magnetic length given in (8.3), namely

$$\upsilon = \frac{N}{A\,B\,e/h\,c} = \frac{N}{\phi/\phi_0} \tag{8.4b}$$

where $\phi = A\,B$ is the magnetic flux through the area A, and $\phi_0 = h\,c/e$ is the flux unit. It is not difficult to prove that the filling factor (8.4) is the quantum number appearing in (8.1). Here we sketch a possible "heuristic" proof of this fact. Let us suppose that the electrons move in the (x, y)-plane with a velocity v_x. Since the magnetic field is directed along the z-axis, the electrons feel a Lorentz force in the y-direction

$$F_y \sim \frac{e}{c} v_x\, B \quad , \tag{8.5}$$

which is then compensated by an emerging electric field E_y, such that

[24] We will show in a moment that the filling factor is the quantum number appearing in (8.1); this is why we used the same symbol.

$$e\,E_y \sim \frac{e}{c} v_x\,B \quad . \tag{8.6}$$

The transverse Hall conductance is defined as

$$\sigma_{\rm H} \equiv \sigma_{xy} = \frac{j_x}{E_y} \tag{8.7}$$

where $j_x \sim e\,\rho\,v_x$ is the electric current in the x-direction. Using (8.6), we obtain

$$\sigma_{\rm H} = \frac{e\,\rho\,v_x}{v_x\,B/c} = \frac{e\,c}{B}\,\rho \tag{8.8}$$

which is a well-known classical result. Using in (8.8) the quantum mechanical relation (8.4) to express ρ in terms of the filling factor v, we get

$$\sigma_{\rm H} = v\,\frac{e^2}{h}$$

as in (8.1). We refer the reader to (Prange and Girvin 1990) for a rigorous derivation of this formula and a discussion of the fundamental role of impurities.

When v is integer (integer QHE) there is an integer number of Landau levels completely filled, and the phenomenon becomes essentially a direct manifestation of the Landau quantization for "non-interacting" electrons in a magnetic field. Furthermore, the system is incompressible and develops a gap. The situation is more complicated and certainly less obvious in the fractional QHE that results from the condensation of the two-dimensional electron system into a new type of collective ground state driven by the Coulomb repulsion. This is incompressible and exhibits a gap like in the integer QHE, despite the fact that only a fraction of Landau levels is filled.

When the fraction is $v = 1/m$, with m an odd integer, this ground state can be described very accurately by the variational wavefunction proposed by R.B. Laughlin (Laughlin 1983)

$$\psi_m = \mathcal{N}_m \prod_{I<J} (z_i - z_j)^m \; {\rm e}^{-\frac{1}{4\ell_0^2}\sum_I |z_I|^2} \tag{8.9}$$

where z_I is the complex coordinate for the I-th electron and $\mathcal{N}_m$ is a normalization factor. Notice that the wavefunction (8.9) has precisely the same form which we found in Chapter 4 for particles of statistics

$$\nu = m \quad . \tag{8.10}$$

Since m is an odd integer, ψ_m is totally antisymmetric, and so it describes ordinary fermions (or better "super"-fermions if $m = 3, 5, \ldots$).

Before exploring how anyons may enter the stage of the fractional QHE, let us examine some properties of the Laughlin wavefunction (8.9). First of all, the prefactor $\prod_{I<J} (z_I - z_J)^m$ is purely analytic, which means that all particles are in the *lowest* Landau level (see Chapter 6). However, ψ_m is not simply the product of

single particle wavefunctions, but is a complicated superposition of such products. The prefactor is also of the Jastrow type: it has a zero of order m at coincident points ($z_I = z_J$), showing that electrons tend very strongly to repel each other in a way that is appropriate to minimize the Coulomb interaction. In this sense they are called sometimes "super"-electrons (with no reference at all to supersymmetry!). If z_I goes around z_J by an angle $\Delta\varphi$, the wavefunction acquires a phase $e^{im\Delta\varphi}$, as if each particle carried m units of flux.

Many of the phenomenological consequences of the Laughlin wavefunction derive from the so called plasma analogy. This is based on the interpretation of the quantum probability density, $|\psi_m|^2$, as a classical Boltzmann distribution with potential energy V_m, according to

$$|\psi_m|^2 = e^{-V_m} \tag{8.11}$$

where

$$V_m = -2m \sum_{I<J} \ln|z_I - z_J| + \frac{1}{2\ell_0^2} \sum_I |z_I|^2 \quad . \tag{8.12}$$

The logarithmic terms of V_m clearly derive from the prefactor of (8.9), while the last term originates from the exponential [25]. It is well-known that a two-dimensional one-component plasma of particles with charge q moving in a neutralizing background of density ρ, is characterized by a potential

$$-q^2 \sum_{I<J} \ln|z_I - z_J| + \frac{1}{2}\pi\rho q^2 \sum_I |z_I|^2 \tag{8.13}$$

where the first term is due to the Coulomb interaction among the particles, and the second term is due to the interaction of the particles with the neutralizing background. Eq. (8.13) has the same form of (8.12) and so we are naturally led to regard V_m as the potential of a plasma of particles with charge

$$q = \sqrt{2m} \tag{8.14}$$

in a neutralizing background of density

$$\rho = \frac{1}{\pi\ell_0^2 q^2} = \frac{1}{2\pi\ell_0^2}\,\frac{1}{m} \quad . \tag{8.15}$$

Variational arguments suggest the identification of ρ in (8.15) with the electron density of the original problem, and therefore from (8.4) we conclude that the wavefunction ψ_m describes a (circular) droplet of fluid with filling factor

$$\nu = 2\pi\ell_0^2\,\rho = \frac{1}{m} \quad . \tag{8.16}$$

There are very strong arguments which support the identification of ψ_m in (8.9) as the wavefunction for the ground state of the fractional QHE at filling

[25] In (8.12) we have neglected the normalization factor $\mathcal{N}_m$ which would give rise only to a trivial constant shift of V_m.

$v = 1/m$, but essentially the justification is that ψ_m has overlap one (with very high accuracy) with the numerical solutions for the QHE computed for various kinds of repulsive potentials (Laughlin 1983). This shows that ψ_m captures the essential and universal features of the fractional QHE since it is largely insensitive to the specific form of the repulsive interactions among electrons.

Any deviations from the density ρ in (8.15) are accommodated in the system as localized quasi-particle or quasi-hole excitations across a gap. These excitations turn out to have *fractional charge* and *fractional statistics*; thus they are anyons! To prove this, let us begin by recalling how such excitations can be obtained. Let us consider the following ideal situation: in the state described by (8.9) we introduce an infinitesimally thin flux-tube at a point z_α, and then turn on adiabatically the flux ϕ from zero to a final value of one unit (*i.e.* $\phi = 0 \longrightarrow \phi = \pm\phi_0 = \pm h\,c/e$), in such a way that the system remains an (instantaneous) eigenstate of the changing Hamiltonian. Due to Faraday law, the variation of the flux from $\phi = 0$ to $\phi = \pm\phi_0$ gives rise to a (circular) electric field around the point z_α. The particles will then flow inwards or outwards (depending on the sign of ϕ), and a net positive or negative charge will accumulate around z_α. However, since the change of flux by one quantum ϕ_0 can be compensated with a gauge transformation (Wu and Yang 1975), the final state can be considered an excited state of the *original* Hamiltonian. The excitation is a quasi-particle or a quasi-hole and its charge turns out to be $e^* = \mp e/m$. In other words, one can say that a quasi-hole (or a quasi-particle) is formed in the incompressible fluid described by ψ_m as a bubble of size such that a fraction $1/m$ of an electron is removed (or added).

For definiteness let us begin by considering the case of a quasi-hole. It is not so difficult to guess that the wavefunction for the state at filling $v = 1/m$ with a quasi-hole at the point z_α is of the form

$$\psi_m^{+z_\alpha} = \mathcal{N}_+ \prod_I (z_I - z_\alpha)\, \psi_m \tag{8.17}$$

where $\mathcal{N}_+$ is a normalization factor and ψ_m is the ground state wavefunction (8.9). Indeed, when the I-th electron travels along a path surrounding other particles, it sees m units of flux at all z_J with $J \neq I$ where the other electrons are located, but when it travels around the quasi-hole in z_α, it sees only *one* unit of flux. Thus, as expected, the quasi-hole behaves like $1/m$ of an electron and in particular carries $1/m$ of its charge (of course with opposite sign!). These conclusions can also be obtained by applying the plasma analogy directly to (8.17) (Laughlin 1983).

The wavefunction describing a quasi-particle excitation above the ground state ψ_m is more complicated and in a certain sense less obvious than that of the quasi-hole. However, also in this case it is possible to write the wavefunction in analytic form (Laughlin 1983), namely

$$\psi_m^{-z_\alpha} = \mathcal{N}_- \prod_I \left(2\ell_0^2 \frac{\partial}{\partial z_I} - \bar{z}_\alpha \right) \psi_m \quad . \tag{8.18}$$

Here ℓ_0 is the magnetic length (8.3) and it is understood that all derivative operators do not act on the exponential factor of ψ_m. Considerations similar to those we have mentioned above for the quasi-hole, allow to conclude that the quasi-particle has a charge $e^* = -e/m$ (notice that the electron charge is $-e$).

The charge of the excitations of the fractional QHE can be computed also in a more direct but more formal way using the concept of Berry phase (Berry 1984), an approach originally advocated by D.P. Arovas, J.R. Schrieffer and F. Wilczek (Arovas *et al.* 1984). This method is very elegant and will allow us to compute also the statistics of the excitations and conclude that they are indeed anyons (cf also (Forte 1991)). For simplicity we begin by considering the case of the quasi-holes wavefunction (8.17). In the Berry phase approach, the charge of the quasi-hole is determined by calculating the change in the phase of the wavefunction $\psi_m^{+z_\alpha}$ as the quasi-hole position z_α is adiabatically moved around a closed loop encircling a flux ϕ. This phase is then compared and identified with the Aharonov-Bohm phase (see (3.2.44)), and the charge is determined accordingly. Before entering the details of this calculation, let us briefly recall the concept of Berry phase (extensive reviews as well as original research papers on this subject can be found in (Shapere and Wiczek 1989)).

Let us consider a quantum system described by a time-dependent Hamiltonian $H(t)$. If we assume that at a given time t_0 the system is described by an instantaneous eigenstate $\psi_\alpha(t_0)$, satisfying

$$H(t)\,\psi_\alpha(t) = E_\alpha(t)\,\psi_\alpha(t) \quad , \tag{8.19}$$

then in the adiabatic approximation (see for instance (Schiff 1968)) the system at time t_1 will be represented by

$$\Psi_\alpha(t_1) = \exp\left[-\frac{\mathrm{i}}{\hbar}\int_{t_0}^{t_1} dt\; E_\alpha(t) - \mathrm{i}\,\gamma_\alpha\right]\psi_\alpha(t_1) \tag{8.20}$$

where the "unusual" phase γ_α is

$$\gamma_\alpha = -\mathrm{i}\int_{t_0}^{t_1} dt\; \langle\psi_\alpha(t)|\frac{d}{dt}|\psi_\alpha(t)\rangle \quad . \tag{8.21}$$

More generally, if the system at time t_0 is a superposition of instantaneous eigenstates $\psi_\alpha(t)$, *i.e.*

$$\Psi(t_0) = \sum_\alpha a_\alpha\,\psi_\alpha(t_0) \quad , \tag{8.22}$$

then in the adiabatic approximation the wavefunction at time t_1 is

$$\Psi(t_1) = \sum_\alpha \exp\left[-\frac{\mathrm{i}}{\hbar}\int_{t_0}^{t_1} dt\; E_\alpha(t) - \mathrm{i}\,\gamma_\alpha\right] a_\alpha\,\psi_\alpha(t_1) \quad . \tag{8.23}$$

The phase γ_α in (8.21) is a geometric phase whose relevance in physics has been recognized only recently (Berry 1984). In fact it is usually unobservable (Schiff 1968), but in some cases, when the wavefunctions are not single-valued, it has interesting and observable consequences. Anyons, whose wavefunctions are multi-valued (see Chapter 4), are therefore the obvious candidates to test the importance of Berry phase.

Let us now return to our problem and compute the Berry phase for the wavefunction (8.17) under the assumption that the quasi-hole is slowly transported around a closed loop Γ so that z_α becomes a time-dependent parameter and the adiabatic approximation can be used. Since the entire time-dependence of the problem comes solely from z_α, using the explicit expression (8.17), we easily obtain

$$\frac{d}{dt}\psi_m^{+z_\alpha}(t) = \sum_I \frac{d}{dt}\left[\ln(z_I - z_\alpha(t))\right] \psi_m^{+z_\alpha}(t) \ , \tag{8.24}$$

and

$$\gamma = -\mathrm{i} \int_{t_0}^{t_1} dt \ \langle \psi_m^{+z_\alpha}(t)| \sum_I \frac{d}{dt}\left[\ln(z_I - z_\alpha(t))\right] |\psi_m^{+z_\alpha}(t)\rangle \ . \tag{8.25}$$

The integral over t in (8.25) is evaluated from the initial time t_0 when the quasi-hole begins its tour, to the final time t_1 when it returns to its original position after a non trivial loop Γ traversed anticlockwise. We can rewrite (8.25) in a more transparent way by observing that the electron density in the state $\psi_m^{+z_\alpha}$, is (see (5.78))

$$\rho(z) = \langle \psi_m^{+z_\alpha}(t)| \sum_I \delta^{(2)}(z - z_I) \ |\psi_m^{+z_\alpha}(t)\rangle \tag{8.26}$$

so that

$$\begin{aligned} \gamma &= -\mathrm{i} \int_{t_0}^{t_1} dt \int d^2z \ \frac{d}{dt}\left[\ln(z - z_\alpha(t))\right] \rho(z) \\ &= -\mathrm{i} \int d^2z \int_{t_0}^{t_1} dt \ \frac{dz_\alpha(t)}{dt} \ \frac{1}{z_\alpha(t) - z} \ \rho(z) \\ &= -\mathrm{i} \int d^2z \oint_\Gamma dz_\alpha \ \frac{1}{z_\alpha - z} \ \rho(z) \ . \end{aligned} \tag{8.27}$$

Under the reasonable assumption that the density $\rho(z)$ is a regular function (see for example (Arovas 1989)), it is easy to realize using the residue theorem that all points z outside Γ give a vanishing contribution to the contour integral, and thus are left with

$$\begin{aligned} \gamma &= -\mathrm{i} \int_{<\Gamma} d^2z \oint_\Gamma dz_\alpha \ \frac{1}{z_\alpha - z} \ \rho(z) \\ &= 2\pi \int_{<\Gamma} d^2z \ \rho(z) = 2\pi \, N_\Gamma \ . \end{aligned} \tag{8.28}$$

The symbol $\int_{<\Gamma} d^2z$ denotes the integral over the region inside the loop Γ, and N_Γ the average number of electrons that are in there. This concludes the calculation of the Berry phase for the quasi-hole wavefunction (8.17).

In Chapter 3, we have seen that if a particle of charge q is moved along a closed loop Γ encircling a flux ϕ_Γ, then its wave function acquires a phase

$$\exp\left[-\mathrm{i}\,\frac{q\,\phi_\Gamma}{\hbar\, c}\right] \tag{8.29}$$

(see (3.2.44)) due to the Aharonov-Bohm effect (Aharonov and Bohm 1959). If we identify the Berry phase

$$\exp\left[-i\,\gamma\right] \tag{8.30}$$

with the Aharonov-Bohm phase (8.29), we have

$$\gamma = \frac{q\,\phi_\Gamma}{\hbar\,c} = 2\pi\,N_\Gamma \quad . \tag{8.31}$$

On the other hand, using (8.4b) and (8.16), the flux ϕ_Γ turns out to be

$$\phi_\Gamma = \frac{N_\Gamma\,\phi_0}{\upsilon} = \frac{h\,c\,N_\Gamma}{e}\,m \;; \tag{8.32}$$

inserting this into (8.31) and solving for q, we finally obtain

$$q = \frac{h\,c\,N_\Gamma}{\phi_\Gamma} = \frac{1}{m}\,e \quad . \tag{8.33}$$

Thus we have proven that the quasi-holes have charge $q \equiv e^* = e/m$.

The calculation of the charge of a quasi-particle proceeds along the same lines, but particular care must be used in treating the derivative operators which appear in the wavefunction (8.18). Formally we can write

$$\frac{d}{dt}\psi_m^{-z_\alpha}(t) = \sum_I \frac{d}{dt}\left[\ln\left(2\ell_0^2\frac{\partial}{\partial z_I} - \bar{z}_\alpha(t)\right)\right]\;\psi_m^{-z_\alpha}(t) \quad , \tag{8.34}$$

where we have to remember that the derivatives $\partial/\partial z_I$ do not act on the exponential factor inside $\psi_m^{-z_\alpha}$. Upon using (8.34) into (8.21), the Berry phase for a quasi-particle becomes

$$\gamma = -i\int_{t_0}^{t_1} dt\;\langle\psi_m^{-z_\alpha}(t)|\;\sum_I \frac{d}{dt}\left[\ln\left(2\ell_0^2\frac{\partial}{\partial z_I} - \bar{z}_\alpha(t)\right)\right]\;|\psi_m^{-z_\alpha}(t)\rangle \quad . \tag{8.35}$$

We stress that (8.34) and (8.35) are only formal expressions whose precise meaning needs to be clarified and specified in the following. This can be accomplished by defining the multi-particle generalization of the Bargmann-Fock inner product (see for instance (Girvin and Jach 1983)) according to

$$\langle\langle g|\;\mathcal{O}\left(\{2\ell_0^2\frac{\partial}{\partial z_I};z_I\}\right)|f\rangle\rangle \equiv$$
$$\int\prod_I d^2z_I\;e^{-\frac{1}{2\ell_0^2}\sum_I |z_I|^2}\;\bar{g}(\{z_I\})\;\mathcal{O}\left(\{2\ell_0^2\frac{\partial}{\partial z_I};z_I\}\right)f(\{z_I\}) \quad . \tag{8.36}$$

Here f and g are two arbitrary analytic functions of many variables, and $\mathcal{O}$ is an operator, functional of $2\ell_0^2\partial/\partial z_I$ and z_I, acting on the space of analytic functions. If such operator is normal-ordered, that is if all derivatives $\partial/\partial z_I$ are put on the left of all complex coordinates z_I, then it is not difficult to prove the following identity (Girvin and Jach 1983)

$$\langle\langle g|\, \mathcal{O}\left(\{2\ell_0^2\frac{\partial}{\partial z_I}; z_I\}\right) |f\rangle\rangle = \langle\langle g|\, \mathcal{O}\left(\{\bar{z}_I; z_I\}\right) |f\rangle\rangle \ . \tag{8.37}$$

Indeed, any derivative $2\ell_0^2\partial/\partial z_I$ acting from the right on the measure factor

$$\mathrm{e}^{-\frac{1}{2\ell_0^2}\sum_I |z_I|^2}$$

of (8.36) yields simply $\bar{z}_I$, and therefore inside the Bargmann-Fock inner product we can safely perform the replacement

$$2\ell_0^2\frac{\partial}{\partial z_I} \longrightarrow \bar{z}_I \ . \tag{8.38}$$

Using the definition (8.36) and remembering that the derivatives $\partial/\partial z_I$ do not act on the exponential factors, it is easy to realize that the integrand of (8.35) can be rewritten as follows

$$\langle\langle \hat{\psi}_m^{-z_\alpha}(t)| \sum_I \frac{d}{dt}\left[\ln\left(2\ell_0^2\frac{\partial}{\partial z_I} - \bar{z}_\alpha(t)\right)\right] |\hat{\psi}_m^{-z_\alpha}(t)\rangle\rangle \tag{8.39}$$

where $\hat{\psi}_m^{-z_\alpha}$ is the wavefunction of the quasi-particle without the exponential factor, that is

$$\hat{\psi}_m^{-z_\alpha}(t) = \mathrm{e}^{\frac{1}{4\ell_0^2}\sum_I |z_I|^2} \quad \psi_m^{-z_\alpha}(t) \ . \tag{8.40}$$

Since the formal operator

$$\sum_I \frac{d}{dt}\left[\ln(2\ell_0^2\frac{\partial}{\partial z_I} - \bar{z}_\alpha(t))\right]$$

is normal-ordered according to our previous definition, upon using (8.38), the Berry phase becomes

$$\gamma = -\mathrm{i}\int_{t_0}^{t_1} dt\, \langle\langle \hat{\psi}_m^{-z_\alpha}(t)| \sum_I \frac{d}{dt}\left[\ln(\bar{z}_I - \bar{z}_\alpha(t))\right] |\hat{\psi}_m^{-z_\alpha}(t)\rangle\rangle \ . \tag{8.41}$$

As in the case of a quasi-hole, γ can be written in a more transparent way using the electron density $\rho(z)$, so that

$$\begin{aligned}
\gamma &= -\mathrm{i}\int_{t_0}^{t_1} dt \int d^2 z\, \frac{d}{dt}\left[\ln(\bar{z} - \bar{z}_\alpha(t))\right] \rho(z) \\
&= -\mathrm{i}\int d^2 z \int_{t_0}^{t_1} dt\, \frac{d\bar{z}_\alpha(t)}{dt}\, \frac{1}{\bar{z}_\alpha(t) - \bar{z}}\, \rho(z) \\
&= -\mathrm{i}\int d^2 z \oint_\Gamma d\bar{z}_\alpha\, \frac{1}{\bar{z}_\alpha - \bar{z}}\, \rho(z) \\
&= -2\pi \int_{<\Gamma} d^2 z\, \rho(z) = -2\pi\, N_\Gamma \ .
\end{aligned} \tag{8.42}$$

Comparing this with the Aharonov-Bohm phase (8.29), we deduce that the quasi-particle have charge

$$q = e^* = -\frac{e}{m} \tag{8.43}$$

as expected.

As we have seen, the calculation of the charge of the excitations above the ground state ψ_m presents essentially no difficulty, and both the plasma analogy and the Berry phase approach give the same results. On the contrary, the issue of what statistics these excitations obey is more subtle, and indeed all possible statistics (fermionic (Laughlin 1983), bosonic (Haldane 1983) and anyonic (Halperin 1984)) have been proposed. In the following we are going to show that the quasi-particles and quasi-holes are anyons of statistics $\nu = 1/m$; the calculation is once again based on the Berry phase γ (Arovas *et al.* 1984). For simplicity and to avoid repetitions, we will examine only the case of quasi-holes.

Let us then consider a state characterized by the presence of two quasi-hole excitations, one located in the point z_α and the other located in the point z_β. If we neglect any interaction between the two quasi-holes, the wavefunction turns out to be

$$\psi_m^{+z_\alpha,+z_\beta} = \mathcal{N}_{\alpha\beta} \prod_I (z_I - z_\alpha)(z_I - z_\beta)\ \psi_m \tag{8.44}$$

which is clearly a straighforward generalization of (8.17). In (8.44) $\mathcal{N}_{\alpha\beta}$ is a normalization factor that is irrelevant for our present discussion. Now let us assume that the quasi-hole in z_α is adiabatically moved on a full loop Γ, while the quasi-hole in z_β is kept fixed. During this process the wavefunction $\psi_m^{+z_\alpha,+z_\beta}$ acquires a Berry phase γ given in (8.21) that can be calculated as before. Skipping the intermediate steps which are exactly the same as in (8.24-27), we have

$$\begin{aligned} \gamma &= -\mathrm{i} \int_{t_0}^{t_1} dt\ \langle \psi_m^{+z_\alpha,+z_\beta}(t)| \frac{d}{dt} |\psi_m^{+z_\alpha,+z_\beta}(t)\rangle \\ &= 2\pi \int_{<\Gamma} d^2z\ \rho(z) \end{aligned} \tag{8.45}$$

where $\rho(z)$ is the electron density in the two quasi-hole state (8.44), namely

$$\rho(z) = \langle \psi_m^{+z_\alpha,+z_\beta}(t)| \sum_I \delta^{(2)}(z - z_I)\ |\psi_m^{+z_\alpha,+z_\beta}(t)\rangle\ . \tag{8.46}$$

If in its motion z_α does not encircle z_β, the integral in (8.45) has the same value as in (8.28) so that

$$\gamma = 2\pi\, N_\Gamma\ . \tag{8.47}$$

Of course this case corresponds to the situation in which the two quasi-holes are not exchanged. Indeed, as far as the quasi-hole configuration space is concerned, the loop Γ is homotopically trivial, according to our general discussion of Chapter 2, because z_α does not encircle z_β. However, if the loop Γ traversed by z_α contains z_β and it is homotopically non-trivial, we have

$$\gamma = 2\pi \int_{<\Gamma} d^2z \; \rho(z) = 2\pi \left(N_\Gamma - \frac{1}{m} \right) \; . \tag{8.48}$$

In this case in fact we have to take into account that a quasi-hole is present inside Γ and hence the number of electrons N_Γ has to be *diminished* by the fraction $1/m$ that is needed to build the quasi-hole. Comparing (8.48) and (8.47) we see that when z_α encircles z_β the wavefunctions picks up an extra phase

$$\exp\left[-\mathrm{i}\,\Delta\gamma\right] = \exp\left[2\pi\mathrm{i}\,\frac{1}{m}\right] \tag{8.49}$$

which can be interpreted as a statistical effect. Indeed in this case the two quasi-holes are exchanged twice, and from (2.1) with $\Delta\varphi = 2\pi$ we conclude that their statistics is

$$\nu = \frac{1}{m} \; . \tag{8.50}$$

When $m = 1$, *i.e.* when the lowest Landau level is completely filled (see (8.16)), the quasi-holes are fermions, but in general for $m = 3, 5, \ldots$ the quasi-holes are anyons of fractional statistics. A similar result can be obtained also for the quasi-particles using the replacement (8.38) inside the expectation values which express the Berry phases.

The derivation of (8.50) clearly shows that the statistics ν is directly related to the fraction of electrons which forms a quasi-hole or a quasi-particle. As we have mentioned, this fraction is a well-defined number as a consequence of the incompressibility of the ground state ψ_m (Laughlin 1983); in particular it does not depend on the details of the loop Γ but only on its homotopy class so that (8.49) can be interpreted as a statistical effect. Furthermore, comparing (8.16) and (8.50), we observe that the excitations above the Laughlin ground state are characterized by a statistics which is numerically equal to the filling factor of ψ_m. This observation will be helpful in deriving the so-called hierarchy scheme (Halperin 1984) which is a systematic way to understand the observed fractions in the QHE and is based on the explicit use of anyons.

We have seen that the quasi-hole excitations above ψ_m have charge $e^* = e/m$ and statistics $\nu = 1/m$; therefore it is natural to expect that the effective wavefunction for two such quasi-holes with coordinates z_α and z_β should contain terms like

$$\psi^+_{m_1} \sim (z_\alpha - z_\beta)^{m_1} \mathrm{e}^{-\frac{1}{4m\ell_0^2}\left(|z_\alpha|^2 + |z_\beta|^2\right)} \tag{8.51}$$

where

$$m_1 = \frac{1}{m} + 2\,p_1 \tag{8.52}$$

with p_1 a positive integer. In fact, under the exchange of the two quasi-holes the factor $(z_\alpha - z_\beta)^{m_1}$ produces the required phase

$$\exp[\mathrm{i}\pi m_1] = \exp\left[\mathrm{i}\pi(\frac{1}{m} + 2\,p_1)\right] = \exp\left[\mathrm{i}\pi\frac{1}{m}\right] \; .$$

The integer p_1 has no effect on the statistics since the latter is defined only modulo 2 and when $p_1 \neq 0$, the quasi-holes of ψ_{m_1} are called "super-anyons" in analogy with the electrons of ψ_m which are sometimes called "super-fermions" when $m = 3, 5, \ldots$. The exponential term in (8.51) is typical for objects of charge $e^* = e/m$, interacting with a magnetic field field B; indeed the magnetic length for the quasi-holes with charge e^* is

$$\ell_0^* = \sqrt{\frac{\hbar c}{e^* B}} = \sqrt{m}\,\ell_0 \tag{8.53}$$

where ℓ_0 is the electron magnetic length (see (8.3)). The effective wavefunction (8.51) for the quasi-holes should result from the wavefunction (8.44) after the integration of all coordinates z_I of the underlying electron system, and can be easily generalized to an arbitrary number of quasi-holes according to

$$\psi_{m_1}^{+} = \tilde{\mathcal{N}}_{+} \prod_{\alpha<\beta} (z_\alpha - z_\beta)^{m_1} \mathrm{e}^{-\frac{1}{4m\ell_0^2}\sum_\alpha |z_\alpha|^2} \ . \tag{8.54}$$

This is clearly the straighforward generalization of the Laughlin wavefunction ψ_m which described an incompressible system of electrons with charge $-e$, statistics $\nu = 1$ and filling factor $v = 1/m$. Using the same arguments, it is therefore easy to conclude that $\psi_{m_1}^{+}$ in (8.54) describes an incompressible system of quasi-holes with charge $e^* = e/m$ and statistics $\nu = 1/m$. Let us now compute the new filling factor by applying the plasma analogy to $\psi_{m_1}^{+}$. We interpret $|\psi_{m_1}^{+}|^2$ as a classical Boltzmann distribution with potential

$$V_{m_1} = -2m_1 \sum_{\alpha<\beta} \ln|z_\alpha - z_\beta| + \frac{1}{2m\ell_0^2} \sum_\alpha |z_\alpha|^2 \ . \tag{8.55}$$

Comparing this expression with (8.13) we immediately find that the density of the neutralizing background is

$$\rho^* = \frac{1}{2\pi\ell_0^2} \frac{1}{m\,m_1} \ . \tag{8.56}$$

As for the electrons, variational arguments again indicate that ρ^* in (8.56) has to be identified with the density of quasi-holes; therefore in an area $2\pi\ell_0^2$ there are

$$2\pi\ell_0^2\,\rho^* = \frac{1}{m\,m_1}$$

quasi-holes carrying a total charge

$$q_1 = \frac{e}{m}\left(\frac{1}{m\,m_1}\right) \ . \tag{8.57}$$

This charge has to be removed from the underlying electron system; hence the *electron* filling factor changes from $v = 1/m$ to

$$\begin{aligned} v_1 &= v - \frac{q_1}{e} \\ &= \frac{1}{m} - \frac{1}{m^2 m_1} \ . \end{aligned} \tag{8.58}$$

Notice that $-q_1/e$ is precisely the number of electrons required to produce a charge q_1 in the area $2\pi\ell_0^2$. Elementary algebra leads to

$$v_1 = \frac{2p_1}{1+2p_1\, m} = \frac{1}{m + \cfrac{1}{2p_1}} \quad . \tag{8.59}$$

We stress once again that this is the electron filling factor resulting from the formation of an incompressible system of quasi-hole excitations above the ground state at filling $1/m$. It is easy to see that if $m = 3$ and $p = 1$, then

$$v_1 = \frac{2}{7}$$

which is one of the experimentally observed fractions in the QHE.

Notice however that the wavefunction $\psi^+_{m_1}$ with $m = 3$ and $p_1 = 1$ does *not* describe the electron state at filling $v_1 = 2/7$ since it is a wavefunction for quasi-holes! However it describes a new incompressible quasi-hole fluid with its definite filling factor which in turn is related to the effective filling factor of the parent electron system. Moreover the fractional statistics of the quasi-holes is in correspondence with this effective filling factor in the same way as the ground state filling factor $v = 1/m$ (with m an odd integer) was related to the fermi statistics of the electrons.

This construction and these observations can be further generalized and extended in a hierarchy scheme (Halperin 1984; Haldane 1983) by considering excitations over excitations in a recursive way. For example one can construct second-generation quasi-hole excitations from the quasi-hole wavefunction $\psi^+_{m_1}$. These are characterized by a charge $e^{**} = -e^*/m_1 = -e/(mm_1)$, and by a statistics $\nu_1 = 1/m_1$. Thus their effective wavefunction would have the form

$$\psi^{++}_{m_2} = \tilde{\mathcal{N}}_{++} \prod_{\alpha<\beta} (z_\alpha - z_\beta)^{m_2} \mathrm{e}^{-\frac{1}{4m\, m_1 \ell_0^2} \sum_\alpha |z_\alpha|^2} \tag{8.60}$$

where

$$m_2 = \frac{1}{m_1} + 2p_2 \tag{8.61}$$

with p_2 being a positive integer. By repeating the same steps as before, one may compute the net filling factor for the electron system: this is given by the original one, $v = 1/m$, diminished by the proper amounts necessary to produce the two generations of quasi-holes, namely

$$v_2 = v_1 - \frac{1}{m\, m_1} \frac{1}{m\, m_1\, m_2} = \left(\frac{1}{m} - \frac{1}{m^2\, m_1}\right) - \frac{1}{m^2\, m_1^2\, m_2} \quad . \tag{8.62}$$

Elementary algebra leads to

$$v_2 = \cfrac{1}{m + \cfrac{1}{2p_1 + \cfrac{1}{2p_2}}} \; . \tag{8.63}$$

Clearly this hierarchy construction can be further iterated, and when also the quasi-particle excitations are taken into account, the generic filling factor is represented by a continued fraction (Haldane 1983)

$$v = \cfrac{1}{m + \cfrac{\alpha_1}{2p_1 + \cfrac{\alpha_2}{2p_2 + \cfrac{\alpha_3}{2p_3 + \ldots}}}} \; . \tag{8.64}$$

Here $\alpha_i = +1$ if the i-th generation consists of quasi-holes (with respect to their parent fluid) and $\alpha_i = -1$ if the i-th generation consists of quasi-particles. In this way all observed fractions are reproduced. Unfortunately however, this hierarchy construction predicts also fractions which are not experimentally observed. To avoid this problem, other approaches have been proposed (Jain 1989a,b, 1990) and we refer the reader to the original literature for their discussion since they do not make an explicit use of the idea of anyons.

9. Anyons and Conformal Field Theories

In the past few years, two-dimensional conformal field theories (CFT) (Belavin *et al.* 1984) have attracted an enormous interest in many different contexts. In fact, CFT describe the scaling limit at the critical point of many statistical mechanical systems in two dimensions (for reviews see for instance (Ginsparg 1990; Cardy 1990)); they are relevant in string theory because they define the background geometry in which the string propagates (see for example (Green *et al.* 1987)); they are essential in the non-perturbative formulation of two-dimensional gravity provided by matrix models (for reviews see (Bilal 1990; Ginsparg 1991)), and last but not least, CFT have a very rich and elegant mathematical structure (Moore and Seiberg 1989). The latter derives essentially from the fact that CFT are quantum theories which carry a non trivial representation of the two-dimensional conformal algebra. This is known also as Virasoro algebra, and reads as follows

$$[L_n\,,\,L_m] = (n-m)\,L_{n+m} + \frac{c}{12}\,(n^3-n)\,\delta_{n+m,0}\quad , \tag{9.1}$$

where n, m are integers and c is the so-called central charge. For a discussion of the properties of the Virasoro algebra and of its representations, we refer the reader to the extensive literature on the subject (see for example (Itzykson *et al.* 1988) where the most relevant papers on CFT are collected). Here we simply recall that in a CFT the N-point correlators of fields describing local observables, are real functions of the complex coordinates where the fields are located, but can be decomposed into a sum of factorized terms, each one being a product of a purely holomorphic factor times a purely anti-holomorphic one. These factors are called conformal blocks and carry non-trivial representations of the braid group B_N (Moore and Seiberg 1989).

The appearance of the braid group naturally leads to conjecture some correspondence between anyons and CFT. A first signal of such correspondence already appeared in Chapter 3, where we pointed out the presence in anyonic systems of the conformal symmetry generated by $L_{\pm 1}$ and L_0, which indeed close the finite sub-algebra of (9.1).

In this chapter we will elaborate on the formal analogy existing between anyons and CFT, and show that the anyonic wavefunctions described in Chapter 6 – in particular those which are relevant for the fractional quantum Hall effect – can be represented as conformal blocks of a particular two-dimensional CFT. The literature on this subject is quite extensive (Fubini 1991; Stone 1991a; Fubini and Lütken 1991; Moore and Read 1991; Cristofano *et al.* 1991a,b,c; Balatsky 1991; Dunne *et al.* 1991a,b; Balatsky and Stone 1991; Nagao 1992a,b; Ting and Lai

1992; Cappelli *et al.* 1992) but still many important and fundamental issues have to be elucidated, in order to really understand the significance of this analogy. In particular it would be nice to exploit the vast knowledge reached in CFT to make further progress in the solution of the anyonic theories, and also to understand the incompressibility of the Laughlin wavefunctions at rational filling factors (see Chapter 8) as a direct consequence of the conformal symmetry.

To be specific, let us consider the case of N identical and indistinguishable particles interacting with an external magnetic field and confined in the lowest Landau level. The single-particle wavefunctions are given by (6.1.12) with $n = 0$, *i.e.*

$$\psi_0^j \sim z^j \, \mathrm{e}^{-\frac{1}{4\ell_0^2}|z|^2} \tag{9.2}$$

where ℓ_0 is the magnetic length (cf (8.3)) and $j = 0, 1, 2, \ldots$ denotes the angular momentum of ψ_0^j. From now on, we will remove the exponential factor from the wavefunctions and absorb it into the integration measure whenever we have to compute matrix elements or overlap integrals. Thus, the "reduced" wavefunctions for the lowest Landau level are simply the powers z^j. To discuss the N-particle states we use the formalism of second quantization under the assumption that the dynamical variables of the single-particle states suffice also to describe a collection of many identical particles, even when these interact with an external field or among themselves (including the case of statistical interactions). Thus, for each single-particle mode z^j with $j = 1, 2, \ldots$ [26], we introduce a creation operator $\hat{\alpha}_j^\dagger$, and an annihilation operator $\hat{\alpha}_j$, satisfying the Heisenberg algebra

$$[\hat{\alpha}_j \, , \, \hat{\alpha}_k^\dagger] = j \; \delta_{j,k} \quad , \tag{9.3}$$

with all other possible commutators being zero. Then, we define a *bosonic* quantum operator $Q(z)$, functional of the single-particle modes, as follows

$$Q(z) = Q_0(z) + \mathrm{i} \sum_{j \neq 0} \frac{\hat{\alpha}_j^\dagger}{j} \; z^j \tag{9.4}$$

where

$$\hat{\alpha}_j^\dagger \equiv \hat{\alpha}_{-j} \qquad \text{for } \; j < 0 \; . \tag{9.5}$$

$Q_0(z)$ in (9.4) is a zero-mode part which we specify in a moment. Since $\hat{\alpha}_j^\dagger$ is a creation operator for $j > 0$ and an annihilation operator for $j < 0$, the Fock vacuum $|0\rangle$ is such that

$$\hat{\alpha}_j^\dagger \, |0\rangle = 0 \qquad \text{for } \; j < 0 \; , \tag{9.6a}$$

and

$$\langle 0| \, \hat{\alpha}_j^\dagger = 0 \qquad \text{for } \; j > 0 \; . \tag{9.6b}$$

[26] The state with $j = 0$ has the lowest angular momentum and can be regarded as the "vacuum" from which all other states are generated by step-operators. Thus, the state with $j = 0$ is treated differently from all other single particle states.

On the other hand, $Q(z)$ is meant to be a quantum bosonic scalar field in two dimensions, and hence its propagator should be

$$\langle 0|\, Q(z)\, Q(w)\, |0\rangle \simeq \ln(z-w) \quad . \tag{9.7}$$

This requirement fixes $Q_0(z)$ to be

$$Q_0(z) = \hat{x} - \mathrm{i}\,\hat{p}\,\ln z \tag{9.8}$$

where $\hat{x}$ and $\hat{p}$ are zero-modes which satisfy

$$[\hat{x}\,,\,\hat{p}] = \mathrm{i} \quad , \tag{9.9}$$

and

$$\hat{p}\,|0\rangle = 0 \quad , \quad \langle 0|\,\hat{x} = 0 \quad . \tag{9.10}$$

Eq. (9.10) implies that $\hat{p}$ can be considered an annihilation operator, and $\hat{x}$ a creation operator on the Fock vacuum $|0\rangle$.

The bosonic field $Q(z)$ given by (9.4) and (9.8), is used in the operator formalism of string theory to describe the string coordinates (for a review see (Green *et al.* 1986)), and also defines a CFT with central charge $c = 1$ whose Virasoro generators are given by

$$L_n = \frac{1}{2} \sum_{j=-\infty}^{\infty} \,:\, \hat{\alpha}^\dagger_{n-j}\,\hat{\alpha}^\dagger_j \,:\quad . \tag{9.11}$$

In this formula, $\hat{\alpha}_0 = \hat{\alpha}^\dagger_0 \equiv \hat{p}$, and the symbol : : denotes the normal ordering in which, as usual, the creation operators are put on the left and the annihilation operators on the right.

Now we wish to show how to use $Q(z)$ to recover the multi-anyon wavefunctions. To this end, we have to keep in mind that the braid group properties of the anyonic wavefunctions cannot be enforced by hand, contrary to the bosonic and fermionic cases where the appropriate symmetrizations or antisymmetrizations can always be performed without any problem. Thus, the information about the anyonic statistics has to be built in the second quantized formalism itself. However, the field $Q(z)$ defined by (9.4) and (9.8) is bosonic in nature and cannot directly be the relevant operator to discuss anyonic statistics. Furthermore, a simple product of N fields $Q(z_I)$ is certainly inadequate to reproduce the anyonic wavefunctions [27] which, as we have seen in Chapter 6, are characterized by a prefactor $\prod\limits_{I<J} (z_I - z_J)^\nu$. These difficulties can be easily overcome if we introduce the so-called Fubini-Veneziano vertex operators which are essentially the exponentials of $Q(z)$ (Fubini and Veneziano 1970). More precisely, to describe anyons of statistics ν we define the vertex operator

$$V_\nu(z) \equiv \,:\, \mathrm{e}^{\mathrm{i}\,\sqrt{\nu}\,Q(z)} \,:\quad , \tag{9.12}$$

[27] To avoid repetitions, we limit ourselves to discuss Type-I wavefunctions. The Type-II wavefunctions can be treated in a similar way if $z \leftrightarrow \bar{z}$ and $\nu \leftrightarrow 2-\nu$.

where : : denotes the normal ordering also of the zero-modes (*i.e.* $\hat{x}$ is put on the left and $\hat{p}$ on the right). If we insert in (9.12) the definition of $Q(z)$ and perform explicitly the normal ordering, we find

$$V_\nu(z) = \exp\left(\sqrt{\nu}\sum_{j=1}^{\infty}\frac{\hat{\alpha}_j^\dagger}{j}\,z^j\right)\ \exp(\mathrm{i}\,\sqrt{\nu}\,\hat{x})\ z^{\sqrt{\nu}\,\hat{p}}\ \exp\left(-\sqrt{\nu}\sum_{j=1}^{\infty}\frac{\hat{\alpha}_j}{j}\,z^{-j}\right)\ . \tag{9.13}$$

The Fubini-Veneziano vertex operators have been extensively used in string theory to compute string scattering amplitudes (Green *et al.* 1987), and in CFT to compute conformal blocks (Ginsparg 1990). Here, following (Fubini 1991; Fubini and Lütken 1991; Dunne *et al.* 1991a,b) we show that the vertex operators can be used also to compute wavefunctions for anyons in the lowest Landau level.

First we note that by a simple reordering, the product of N vertex operators can be rewritten as

$$\begin{aligned}\prod_{I=1}^{N} V_\nu(z_I) = \prod_{I<J}(z_I - z_J)^\nu\ &\exp\left[\sqrt{\nu}\sum_{j=1}^{\infty}\frac{\hat{\alpha}_j^\dagger}{j}\left(\sum_{I=1}^{N} z_I^j\right)\right] \\ \times \exp(\mathrm{i}\,\sqrt{\nu}\,N\,\hat{x})\ \prod_{I=1}^{N} z_I^{\sqrt{\nu}\,\hat{p}}\ &\exp\left[-\sqrt{\nu}\sum_{j=1}^{\infty}\frac{\hat{\alpha}_j}{j}\left(\sum_{I=1}^{N} z_I^j\right)\right]\ .\end{aligned} \tag{9.14}$$

This has the immediate consequence that the vertex operators are good candidates for anyonic operators in that they satisfy anyonic exchange properties. In fact, one has

$$V_\nu(z_1)V_\nu(z_2) = V_\nu(z_2)V_\nu(z_1)\ e^{\pm\mathrm{i}\,\nu\,\pi}\ , \tag{9.15}$$

where the signs in the exponent correspond to the two topologically distinct ways of interchanging z_1 and z_2 (*i.e.* clockwise and counterclockwise). Moreover, the vertex operator (9.12) is a primary field of weight $\nu/2$ in the CFT whose Virasoro generators are given in (9.11); this means that

$$[L_n\,,\,V_\nu(z)] = z^n\left(z\,\frac{d}{dz} + \frac{\nu}{2}\,(n+1)\right)V_\nu(z) \tag{9.16}$$

for any n. In particular, (9.16) for $n=0$ implies that the vertex operator behaves as a fractional spin operator under rotations, namely

$$\mathrm{e}^{\mathrm{i}\,L_0\,\theta}\ V_\nu(z)\ \mathrm{e}^{-\mathrm{i}\,L_0\,\theta} = V_\nu(z\,\mathrm{e}^{\mathrm{i}\,\theta})\ \mathrm{e}^{\mathrm{i}\,\frac{\nu}{2}\,\theta}\ , \tag{9.17}$$

where the rotation generator L_0 has the oscillator representation given in (9.11) and θ is the rotation angle. From (9.17) we can read the value of the spin that is

$$s = \frac{\nu}{2}\ , \tag{9.18}$$

in agreement with the spin-statistics connection.

Now we wish to show that the expectation values of products of vertex operators, which in CFT represent conformal blocks, coincide with the reduced anyon wavefunctions. To this end, let us define the state $|p\rangle$ and its dual $\langle\langle p|$ such that

$$\begin{aligned} \hat{p}\ |p\rangle &= p\ |p\rangle \quad , \quad \langle\langle p|\ \hat{p} = \langle\langle p|\ p\ , \\ \hat{\alpha}_j^{\dagger}\ |p\rangle &= 0 \qquad \text{for } j<0\ , \\ \langle\langle p|\ \hat{\alpha}_j^{\dagger} &= 0 \qquad \text{for } j>0\ , \\ \langle\langle p|p'\rangle &= \delta(p-p')\ . \end{aligned} \tag{9.19}$$

Using (9.9), it is straightforward to prove that

$$(\mathrm{i}\,\hat{x})\ |p\rangle = |p+1\rangle \quad , \quad \langle\langle p|\ (\mathrm{i}\,\hat{x}) = \langle\langle p-1|\ . \tag{9.20}$$

Clearly the state $|p=0\rangle$ is the Fock vacuum $|0\rangle$ (cf (9.10)), but the state $\langle\langle p=0|$ is *not* the dual vacuum $\langle 0|$. In fact, the latter is annihilated by $\hat{x}$, while $\langle\langle p=0|$ is annihilated by $\hat{p}$. However, we can easily establish the precise relationship between $\langle 0|$ and the states $\langle\langle p|$ if we interpret $\hat{x}$ and $\hat{p}$ as a standard pair of canonically conjugate operators [28]. Let us define the state $\langle x|$ such that

$$\langle x|\ \hat{x} = \langle x|\ x\ ,$$

then using the Fourier transformation and going to the momentum representation, we can write

$$\langle x| = \sum_p \langle\langle p|\ \mathrm{e}^{-\mathrm{i}px} \tag{9.21}$$

where the sum $\sum_p$ is over all the allowed values of the momentum p. Putting $x=0$ in this equation, we find

$$\langle 0| = \sum_p \langle\langle p|\ . \tag{9.22}$$

Thus, the dual Fock vacuum $\langle 0|$ is a combination of all dual states with definite momentum (this is of course a natural consequence of (9.9) and (9.10)). Let us now compute the following expectation value

$$\langle\langle p|\ \prod_{I=1}^{N} V_{\nu}(z_I)\ |0\rangle\ .$$

Using (9.14) and (9.19), we obtain

$$\langle\langle p|\ \prod_{I=1}^{N} V_{\nu}(z_I)\ |0\rangle = \prod_{I<J} (z_I - z_J)^{\nu}\ \delta(p - N\sqrt{\nu}) \tag{9.23}$$

[28] We recall that in string theory $\hat{x}$ and $\hat{p}$ are respectively the position and the momentum of the center of mass of the string.

where the δ-function arises from momentum conservation. If we make use of (9.22), our result can be written also as follows

$$\langle 0| \prod_{I=1}^{N} V_\nu(z_I)\ |0\rangle = \prod_{I<J} (z_I - z_J)^\nu \ . \tag{9.24}$$

Notice that momentum conservation selects the momentum state $\langle\langle\sqrt{\nu}\,N|$ from the dual vacuum $\langle 0|$. Thus, the vacuum expectation value of a product of vertex operators is precisely the statistical prefactor which characterizes anyonic wavefunctions.

We can proceed even further and use this operator formalism to reproduce the entire wavefunctions for anyons in the lowest Landau level, and not just their statistical prefactors. In the dual Fock space of the Q field, let us consider the state

$$\langle\{\lambda_k\}| = \langle 0| \left(\frac{\hat{\alpha}_1}{\sqrt{\nu}}\right)^{\lambda_1} \left(\frac{\hat{\alpha}_2}{\sqrt{\nu}}\right)^{\lambda_2} \cdots \left(\frac{\hat{\alpha}_k}{\sqrt{\nu}}\right)^{\lambda_k} \ , \tag{9.25}$$

where $\{\lambda_k\}$ is an ordered set of non-negative integers, and let us denote by d its level, which is defined by

$$d \equiv \sum_k k\,\lambda_k \ . \tag{9.26}$$

A straightforward computation then shows that

$$\langle\{\lambda_k\}| \prod_{I=1}^{N} V_\nu(z_I)\ |0\rangle = B^{(d)}_{\{\lambda_k\}}(z_1, \ldots z_N) \prod_{I<J} (z_I - z_J)^\nu \ , \tag{9.27}$$

where $B^{(d)}_{\{\lambda_k\}}(z_1, \ldots z_N)$ is the following symmetric homogeneous polynomial of degree d

$$B^{(d)}_{\{\lambda_k\}}(z_1, \ldots z_N) = \left(\sum_{I_1=1}^{N} z_{I_1}\right)^{\lambda_1} \left(\sum_{I_2=1}^{N} {z_{I_2}}^2\right)^{\lambda_2} \cdots \left(\sum_{I_k=1}^{N} {z_{I_k}}^k\right)^{\lambda_k} \ . \tag{9.28}$$

Apart from the exponential factor that has been removed, (9.27) is precisely the *generic* wavefunction for anyons in the lowest Landau level with angular momentum

$$J = \nu\,\frac{N(N-1)}{2} + d \ .$$

Therefore, the insertion of the oscillators $\hat{\alpha}_k$ in the correlation function (9.24) is seen to increase the angular momentum of the base state $\prod_{I<J} (z_I - z_J)^\nu$. In fact, like in string theory, $\langle\{\lambda_k\}|$ is a higher angular momentum state.

It is interesting to consider some applications of this formalism to describe the wavefunctions of the fractional quantum Hall effect that we discussed in the previous chapter. In particular, it is immediate to realize that the relevant part of the Laughlin wavefunction $\hat{\psi}_m$ given in (8.9) can be written as

$$\hat{\psi}_m = \langle 0|\ \prod_{I=1}^{N} V_m(z_I)\ |0\rangle\ , \tag{9.29}$$

with m being an odd integer. Thus, the vertex operator formalism provides a naturally factorized expression for the Laughlin wavefunction in which the electron located at z_I is represented by the fermionic vertex operator $V_m(z_I)$. Similar considerations apply also to the wavefunction for quasi-hole excitations, whose relevant part is (cf (8.17))

$$\hat{\psi}_m^{+z_\alpha} = \prod_{I=1}^{N}(z_\alpha - z_I)\ \prod_{I<J}^{N}(z_I - z_J)^m\ , \tag{9.30}$$

where z_α is the quasi-hole position. In fact, it is not difficult to see that $\hat{\psi}_m^{+z_\alpha}$ can be written as follows

$$\hat{\psi}_m^{+z_\alpha} = \langle 0|\ V_{\frac{1}{m}}(z_\alpha)\ \prod_{I=1}^{N} V_m(z_I)\ |0\rangle\ . \tag{9.31}$$

This expression identifies the anyonic vertex operator $V_{\frac{1}{m}}(z_\alpha)$ as the one describing a quasi-hole excitation at the point z_α. Notice that using (9.15), it is immediate to see that the statistics of the quasi-hole is $\nu = 1/m$, in agreement with the Berry phase derivation which we described in Chapter 8 (cf (8.50)).

The situation is more complicated for the quasi-particle wavefunctions which are not holomorphic even after the removal of the exponential factor (cf (8.18)). Therefore, to give a representation of these wavefunctions in terms of vertex operators it is necessary to enlarge the previous discussion and include also the anti-holomorphic modes. This procedure is well-known in string theory (Virasoro 1969; Shapiro 1970) where it is used to describe the closed string (Green *et al.* 1987). In particular, one introduces a new set of operators, $\tilde{\alpha}_j^\dagger$, $\tilde{\alpha}_j$, $\tilde{x}$ and $\tilde{p}$, such that

$$\begin{aligned} [\tilde{\alpha}_j\, ,\, \tilde{\alpha}_k^\dagger] &= j\ \delta_{j,k}\ , \\ [\tilde{x}\, ,\, \tilde{p}] &= \mathrm{i}\ . \end{aligned} \tag{9.32}$$

These new operators commute with those we have previously introduced. Then, one defines a new bosonic field $\tilde{Q}(\bar{z})$ which is anti-holomorphic and, correspondingly, a new vertex operator

$$\tilde{V}_\nu(\bar{z}) \equiv\ :\ \mathrm{e}^{\mathrm{i}\sqrt{\nu}\tilde{Q}(\bar{z})}\ :\ . \tag{9.33}$$

Assuming that the states $|0\rangle$ and $\langle 0|$ are Fock vacua also for the new oscillators – *i.e.* that (9.6) and (9.10) hold also for $\tilde{\alpha}_j^\dagger$, $\tilde{\alpha}_j$, $\tilde{x}$ and $\tilde{p}$ – it is easy to derive that

$$\langle 0|\ \prod_{I=1}^{N} V_\nu(z_I)\ :\ \prod_{I=1}^{N} \tilde{V}_\nu(\bar{z}_I)\ :\ |0\rangle = \prod_{I<J}^{N}(z_I - z_J)^\nu\ . \tag{9.34}$$

The normal ordering on the product of the anti-holomorphic vertex operators reduces to 1 their contribution to the vacuum expectation value. Therefore, their use in (9.34) looks like a cumbersome and unnecessary complication. However, the doubling of the operators and (9.34) allow us to represent the quasi-particle wavefunctions of the fractional quantum Hall effect (Fubini and Lütken 1991). To see this, let us put $\nu = m$ and insert in (9.34) a vertex operator $\tilde{V}_{\frac{1}{m}}(\bar{z}_\alpha)$; after some simple algebra we get

$$\langle 0|\, \tilde{V}_{\frac{1}{m}}(\bar{z}_\alpha) \prod_{I=1}^{N} V_\nu(z_I) \,:\, \prod_{I=1}^{N} \tilde{V}_\nu(\bar{z}_I) \,:\, |0\rangle = \prod_{I=1}^{N}(\bar{z}_\alpha - \bar{z}_I) \prod_{I<J}^{N}(z_I - z_J)^\nu \; . \quad (9.35)$$

The right hand side of (9.35) is weakly equivalent to the reduced quasi-particle wavefunction $\hat{\psi}_m^{-z_\alpha}$ given in (8.18). This means that the two expressions are not necessarily identical but yield the same matrix elements which are expressed as Bargmann-Fock inner products (see (8.36)). Indeed, as we discussed in Chapter 8, inside these inner products one can safely perform the replacements

$$2\,\ell_0^2 \,\frac{\partial}{\partial z_I} \leftrightarrow \bar{z}_I \quad ,$$

so that (9.35) when projected onto the lowest Landau level, becomes equivalent to $\hat{\psi}_m^{-z_\alpha}$ (Girvin and Jach 1983). With this interpretation, the vertex operator $\tilde{V}_{\frac{1}{m}}(\bar{z}_\alpha)$ can be regarded as creating a quasi-particle at the point z_α, and upon using the obvious extension of (9.15), it is immediate to realize that the quasi-particle statistics is $\nu = 1/m$ in agreement with the Berry phase calculation of the previous chapter. Furthermore, since $V_{\frac{1}{m}}(z_\alpha)$ represents a quasi-hole and $\tilde{V}_{\frac{1}{m}}(\bar{z}_\alpha)$ represents a quasi-particle, the particle-hole duality (Laughlin 1983) is manifest in this vertex operator formalism suggested by string theory and CFT.

We conclude this chapter with a few comments. In analogy with string theory where the Fubini-Veneziano vertex operators can be used to compute multi-loop scattering amplitudes on Riemann surfaces of higher genus (see for instance (Di Vecchia *et al.* 1989)), here too one can think of using the vertex operators (9.12) and (9.33) to compute the anyon wavefunctions on surfaces of non-trivial topology. Indeed, in (Cristofano *et al.* 1991a,b,c) this has been done for the torus, and the genus one generalization of the Laughlin wavefunctions (Haldane and Rezayi 1985) – as well as their degeneracy – have been obtained in the operator formalism. Preliminary results for higher genus Riemann surfaces are also available (Maharana and Panda 1991).

The discussion we have presented here is limited to the case in which all anyons are in the lowest Landau level. Despite the extensive literature on the subject (Fubini 1991; Stone 1991a; Fubini and Lütken 1991; Cristofano *et al.* 1991a,b,c; Balatsky 1991; Dunne *et al.* 1991a,b; Balatsky and Stone 1991; Nagao 1992a,b; Ting and Lai 1992), it is not known at the moment how to generalize the CFT construction with vertex operators to the case in which some particles are in excited states. We think that this is a necessary step towards a full understanding of the analogy between the conformal blocks of CFT and the anyon wavefunctions. In (Moore and Read 1991) this analogy has been studied in some more detail, and the ultimate reason for its validity has been indicated in the relation of the two-dimensional CFT on one hand, and of the anyon theory on the other, with the Chern-Simons field theory. In fact gauge theories in (2+1) dimensions with Chern-Simons action, are topological field theories which reproduce the braiding and fusing properties of two-dimensional CFT (Witten 1989; Moore and Seiberg 1989). On the other hand, Chern-Simons gauge fields are intimately connected to anyons as we discussed at length in Chapters 3 and 5. This circle of ideas needs to be analyzed and studied in more detail, and perhaps the best place to do so is provided by the gapless edge excitations of the fractional quantum Hall effect (Halperin 1981) where non-abelian generalization of fractional statistics and Kac-Moody algebras seem to play a crucial role (Wen 1990b; Stone 1991b; Fröhlich and Kerler 1991). Finally, it would be quite interesting to understand the role and the meaning of the Virasoro algebra (9.1) in the context of anyon theories, and to clarify the significance of its non-linear extensions (w-algebras) which have been recently identified in quantum Hall systems (Cappelli *et al.* 1992).

These issues are still under current active investigations, and I hope that the material and the discussions presented in this chapter as well as in the rest of this book, might be of some help in these developments.

References

Aharonov and Bohm 1959: Y. Aharonov and D. Bohm, Phys. Rev. **115** (1959) 485

Alvarez-Gaumé and Ginsparg 1985: L. Alvarez-Gaumé and P. Ginsparg, Ann. Phys. **161** (1985) 423; Erratum *ibidem* **171** (1986) 233

Arovas 1989: D.P. Arovas, in *Geometric Phases in Physics* edited by A. Shapere and F. Wilczek (World Scientific, Singapore 1989)

Arovas *et al.* 1984: D.P. Arovas, J.R. Schrieffer and F. Wilczek, Phys. Rev. Lett. **53** (1984) 772

Arovas *et al.* 1985: D.P. Arovas, J.R. Schrieffer, F. Wilczek and A. Zee, Nucl. Phys. **B251** (1985) 117

Artin 1926: E. Artin, Abh. Math. Sem. Hamburg **4** (1926) 47

Artin 1947: E. Artin, Annals of Math. **48** (1947) 101

Avron *et al.* 1983: J.E. Avron, R. Seiler and B. Simon, Phys. Rev. Lett. **51** (1983) 51

Balatsky 1991: A.V. Balatsky, Phys. Rev. **B43** (1991) 1257

Balatsky and Stone 1991: A.V. Balatsky and M. Stone, Phys. Rev. **B43** (1991) 8038

Belavin *et al.* 1984: A.A. Belavin, A.M. Polyakov and A.B. Zamolodchikov, Nucl. Phys. **B241** (1984) 333

Berry 1984: M.V. Berry, Proc. R. Soc. London **A392** (1984) 45

Bhaduri *et al.* 1991: R.K. Bhaduri, R.S. Bhalerao, A. Khare, J. Law and M.V.N. Murthy, Phys. Rev. Lett. **66** (1991) 523

Bilal 1990: A. Bilal, C.E.R.N. preprint, CERN-Th 5867/90 (September 1990)

Birman 1969: J.S. Birman, Comm. Pure Appl. Math. **22** (1969) 41

Birmingham *et al.* 1991: D. Birmingham, M. Blau, M. Rakowski and G. Thompson, Phys. Rep. **209** (1991) 199

Blum *et al.* 1990: T. Blum, C.R. Hagen and S. Ramaswamy, Phys. Rev. Lett. **64** (1990) 709

Cardy 1990: J.L. Cardy, in *Fields, Strings and Critical Phenomena* edited by E. Brézin and J. Zinn-Justin (North Holland, Amsterdam 1990)

Cappelli *et al.* 1992: A. Cappelli, C.A. Trugenberger and G. Zemba, C.E.R.N. preprint, CERN-TH 6516/92 (May 1992)

Ceresole *et al.* 1986: A. Ceresole, A. Lerda, P. van Nieuwenhuizen and P. Pizzochero, in *Proceedings of the Fourth Marcel Grossmann Meeting on General Relativity* edited by R. Ruffini (Elsevier Science Publisher B.V., Amsterdam 1986)

Chen *et al.* 1989: Y.-H. Chen, F. Wilczek, E. Witten and B.I. Halperin, Int. Jour. Mod. Phys. **B3** (1989) 1001

Chin and Hu 1992: S.A. Chin and C.R. Hu, Texas A. & M. preprint, CTP-TAMU-01-92 (January 1992)

Cho and Rim 1992: K.H. Cho and C. Rim, Ann. Phys. **213** (1992) 295

Chou 1991a: C. Chou, Phys. Lett. **155A** (1991) 245

Chou 1991b: C. Chou, Phys. Rev. **D44** (1991) 2533; Erratum *ibidem* **D45** (1992) 1433

Chou *et al.* 1992: C. Chou, L. Hua and G. Amelino-Camelia, M.I.T. preprint, MIT-CTP-2059 (January 1992)

Comtet *et al.* 1989: A. Comtet, Y. Georgelin and S. Ouvry, Jour. Phys. **A22** (1989) 3917

Comtet *et al.* 1991: A. Comtet, J. McCabe and S. Ouvry, Phys. Lett. **260B** (1991) 372

Cristofano *et al.* 1991a: G. Cristofano, G. Maiella, R. Musto and F. Nicodemi, Phys. Lett. **262B** (1991) 88

Cristofano *et al.* 1991b: G. Cristofano, G. Maiella, R. Musto and F. Nicodemi, Mod. Phys. Lett. **A6** (1991) 1779

Cristofano *et al.* 1991c: G. Cristofano, G. Maiella, R. Musto and F. Nicodemi, Mod. Phys. Lett. **A6** (1991) 2985

Dasnières de Veigy and Ouvry 1991: A. Dasnières de Veigy and S. Ouvry, I.P.N. preprint, IPNO-TH-91-96 (December 1991)

Deser *et al.* 1982a: S. Deser, R. Jackiw and S. Templeton, Phys. Rev. Lett. **48** (1982) 975

Deser *et al.* 1982b: S. Deser, R. Jackiw and S. Templeton, Ann. Phys. **140** (1982) 372; Erratum *ibidem* **185** (1988) 406

Dirac 1931: P.A.M. Dirac, Proc. R. Soc. London **A133** (1931) 60

Di Vecchia *et al.* 1989: P. Di Vecchia, F. Pezzella, M. Frau, K. Hornfeck, A. Lerda and S. Sciuto, Nucl. Phys. **B322** (1989) 317

Dunne 1992: G.V. Dunne, M.I.T. preprint, MIT-CTP-2079 (March 1992)

Dunne *et al.* 1991a: G.V. Dunne, A. Lerda and C.A. Trugenberger, Mod. Phys. Lett. **A6** (1991) 2891

Dunne *et al.* 1991b: G.V. Dunne, A. Lerda and C.A. Trugenberger, Int. Jour. Mod. Phys. **B5** (1991) 1675

Dunne *et al.* 1991c: G.V. Dunne, R. Jackiw, S.-Y. Pi and C.A. Trugenberger, Phys. Rev. **D43** (1991) 1332

Dunne *et al.* 1992a: G.V. Dunne, A. Lerda, S. Sciuto and C.A. Trugenberger, Nucl. Phys. **B370** (1992) 601

Dunne *et al.* 1992b: G.V. Dunne, A. Lerda, S. Sciuto and C.A. Trugenberger, Phys. Lett. **277B** (1992) 474

Einarsson 1990: T. Einarsson, Phys. Rev. Lett. **64** (1990) 1995

Einarsson 1991: T. Einarsson, Mod. Phys. Lett. **B5** (1991) 675

Ezawa *et al.* 1991a: Z.F. Ezawa, M. Hotta and A. Iwazaki, Phys. Rev. **D44** (1991) 452

Ezawa *et al.* 1991b: Z.F. Ezawa, M. Hotta and A. Iwazaki, Tohoku University preprint, TU-376 (July 1991)

Ezawa and Iwazaki 1991: Z.F. Ezawa and A. Iwazaki, Phys. Rev. **B43** (1991) 2637

Fadell and Neuwirth 1962: E. Fadell and L. Neuwirth, Math. Scand. **10** (1962) 111

Fadell and Van Buskirk 1962: E. Fadell and J. Van Buskirk, Duke Math. J. **29** (1962) 243

Farkas and Kra 1980: H.M. Farkas and I. Kra, *Riemann Surfaces* (Springer-Verlag, New York 1980)

Feyman and Hibbs 1965: *Quantum Mechanics and Path Integrals* (McGraw-Hill, New York 1965)

Finkelstein and Rubistein 1968: D. Finkelstein and J. Rubinstein, J. Math. Phys. **9** (1968) 1762

Forte 1991: S. Forte, Mod. Phys. Lett. **A6** (1991) 3153

Forte 1992a: S. Forte, Rev. Mod. Phys. **64** (1992) 193

Forte 1992b: S. Forte, Int. Jour. Mod. Phys. **A7** (1992) 1025

Forte and Jolicoeur 1991: S. Forte and T. Jolicoeur, Nucl. Phys. **B350** (1991) 589

Fox and Neuwirth 1962: R. Fox and L. Neuwirth, Math. Scand. **10** (1962) 119
Fröhlich and Marchetti 1988: J. Fröhlich and P.A. Marchetti, Lett. Math. Phys. **16** (1988) 347
Fröhlich and Marchetti 1989: J. Fröhlich and P.A. Marchetti, Comm. Math. Phys. **121** (1989) 177
Fröhlich *et al.* 1989: J. Fröhlich, F. Gabbiani and P.A. Marchetti, in *Knots, Topology and Quantum Field Theories* edited by L. Lusanna (World Scientific, Singapore 1989)
Fröhlich and Kerler 1991: J. Fröhlich and T. Kerler, Nucl. Phys. **B354** (1991) 365
Fubini 1991: S. Fubini, Mod. Phys. Lett. **A6** (1991) 347
Fubini and Lütken 1991: S. Fubini and C.A. Lütken, Mod. Phys. Lett. **A6** (1991) 487
Fubini and Veneziano 1970: S. Fubini and G. Veneziano, Nuovo Cim. **67A** (1970) 29
Georgi 1982: *Lie Algebras in Particle Physics*, Benjamin/Cummings, Reading, MA (1982)
Gerbert 1990: P. de Sousa Gerbert, Phys. Rev. **D42** (1990) 543
Ginsparg 1990: P. Ginsparg, in *Fields, Strings and Critical Phenomena* edited by E. Brézin and J. Zinn-Justin (North Holland, Amsterdam 1990)
Ginsparg 1991: P. Ginsparg, Los Alamos preprint, LA-UR-91-9999 (September 1991)
Girvin and Jach 1983: S.M. Girvin and T. Jach, Phys. Rev. **B29** (1983) 5617
Girvin *et al.* 1990: S.M. Girvin, A.H. Mac Donald, M.P.A. Fisher, S.-J. Rey and J.P. Sethna, Phys. Rev. Lett. **65** (1990) 1671
Goldhaber 1982: A.S. Goldhaber, Phys. Rev. Lett. **49** (1982) 905
Goldhaber and Mackenzie 1988: A.S. Goldhaber and R. Mackenzie, Phys. Lett. **214B** (1988) 471
Goldin *et al.* 1980: G.A. Goldin, R. Menikoff and D.H. Sharp, J. Math. Phys. **21** (1980) 650
Goldin *et al.* 1981: G.A. Goldin, R. Menikoff and D.H. Sharp, J. Math. Phys. **22** (1981) 1664
Goldin and Sharp 1991: G.A. Goldin and D.H. Sharp, Int. Jour. Mod. Phys. **B5** (1991) 2625
Green and Schwarz 1984: M.B. Green and J.H. Schwarz, Phys. Lett. **149B** (1984) 117
Green *et al.* 1987: M.B. Green, J.H. Schwarz and E. Witten, *Superstring Theory*, Cambridge University Press, Cambridge (1987)
Greenberg 1964: O.W. Greenberg, Phys. Rev. Lett. **13** (1964) 598
Grundberg *et al.* 1991a: J. Grundberg, T.H. Hansson, A. Karlhede and J.M. Leinaas, Mod. Phys. Lett. **B5** (1991) 539
Grundberg *et al.* 1991b: J. Grundberg, T.H. Hansson, A. Karlhede and E. Westerberg, Phys. Rev. **B44** (1991) 8373
Hagen 1984: C.R. Hagen, Ann. Phys. **157** (1984) 342
Haldane 1983: F.D.M. Haldane, Phys. Rev. Lett. **51** (1983) 605
Haldane and Rezayi 1985: F.D.M. Haldane and E.H. Rezayi, Phys. Rev. Lett. **54** (1985) 237
Halperin 1981: B.I. Halperin, Phys. Rev. **B25** (1981) 2185
Halperin 1984: B.I. Halperin, Phys. Rev. Lett. **52** (1984) 1583
Halperin *et al.* 1989: B.I. Halperin, J. March-Russell and F. Wilczek, Phys. Rev. **B40** (1989) 8726
Han and Nambu 1965: M. Han and Y. Nambu, Phys. Rev. **139B** (1965) 1006
Hansson *et al.* 1988: T.H. Hansson, M. Roček, I. Zahed and S.C. Zhang, Phys. Lett. **214B** (1988) 475
Hasenfratz and 't Hooft 1976: P. Hasenfratz and G. 't Hooft, Phys. Rev. Lett. **36** (1976) 1119

Huang 1987: K. Huang, *Statistical Mechanics* (John Wiley & Sons Inc., New York 1987)
Iengo and Lechner 1990: R. Iengo and K. Lechner, Nucl. Phys. **B346** (1990) 551
Iengo and Lechner 1991: R. Iengo and K. Lechner, Nucl. Phys. **B364** (1991) 551
Iengo and Lechner 1992: R. Iengo and K. Lechner, Phys. Rep. **213** (1992) 179
Illuminati *et al.* 1992: F. Illuminati, F. Ravndal and J.A. Ruud, Phys. Lett. **161A** (1992)
Imbo and March-Russell 1990: T. Imbo and J. March-Russell, Phys. Lett. **252B** (1990) 84
Itzykson *et al.* 1988: *Conformal Invariance and Applications to Statistical Mechanics* edited by C. Itzykson, H. Saleur and J.-B. Zuber (World Scientific, Singapore 1988)
Jackiw 1990: R. Jackiw, Ann. Phys. **201** (1990) 83
Jackiw and Pi 1990: R. Jackiw and S.-Y. Pi, Phys. Rev. **D42** (1990) 3500
Jackiw and Rebbi 1976: R. Jackiw and C. Rebbi, Phys. Rev. Lett. **36** (1976) 1116
Jackiw and Redlich 1983: R. Jackiw and A.N. Redlich, Phys. Rev. Lett. **50** (1983) 555
Jackiw and Templeton 1981: R. Jackiw and S. Templeton, Phys. Rev. **D23** (1981) 2291
Jain 1989a: J.Jain, Phys. Rev. Lett. **63** (1989) 199
Jain 1989b: J.Jain, Phys. Rev. **B40** (1989) 8079
Jain 1990: J.Jain, Phys. Rev. **B41** (1990) 7653
Johnson and Canright 1990: M.D. Johnson and G.S. Canright, Phys. Rev. **B41** (1990) 6870
Kiefl *et al.* 1990: R.F. Kiefl, J.H. Brewer, I. Affleck, J.F. Carolan, P. Dosanjh, W.N. Hardy, T. Hsu, R. Kadono, J.R. Kempton, S.R. Kreitzman, Q. Li, A.H. O'Reilly, T.M. Riseman, P. Schleger, P.C.E. Stamp, H. Zhou, L.P. Le, G.M. Luke, B. Sternlieb, Y.J. Venmra, H.R. Hart and K.W. Lay, Phys. Rev. Lett. **64** (1990) 2082
Khare and McCabe 1991: A. Khare and J. McCabe, Phys. Lett. **269B** (1991) 330
Karlhede and Westerberg 1991: A. Karlhede and E. Westerberg, University of Stockhoklm preprint, USITP-91-8 (August 1991)
Kogan 1991: I.I. Kogan, Phys. Lett. **262B** (1991) 83
Kohmoto 1985: M. Kohmoto, Ann. Phys. **160** (1985) 355
Laidlaw and Morette-de Witt 1971: M.G.G. Laidlaw and C. Morette-de Witt, Phys. Rev. **D6** (1971) 1375
Landau 1930: L.D. Landau, Zeit. für Phys. **64** (1930) 629
Landau and Lifschitz 1977: *Quantum Mechanics: Non-Relativistic Theory*, Pergamon Press, Oxford (1977)
Laughlin 1983: R.B. Laughlin, Phys. Rev. **B23** (1983) 3383
Leinaas and Myrheim 1977: J.M. Leinaas and J. Myrheim, Nuovo Cim. **37B** (1977) 1
Lykken *et al.* 1991: J.D. Lykken, J. Sonnenschein and N. Weiss, Int. Jour. Mod. Phys. **A6** (1991) 5155
Lyons *et al.* 1990: K.B. Lyons, J. Kwo, J.F. Dillon Jr., G.P. Espinosa, M. McGlashan-Powell, A.P. Ramirez and L.F. Schneemeyers, Phys. Rev. Lett. **64** (1990) 2949
Nambu 1966: Y. Nambu, in *Preludes in Theoretical Physics* edited by A. De Shalit (North Holland, Amsterdam 1966)
Nagao 1992a: G. Nagao, University of Texas at Austin preprint, DOE-287-CPP-25 (February 1992)
Nagao 1992b: G. Nagao, University of Texas at Austin preprint, DOE-288-CPP-26 (February 1992)
Niu *et al.* 1985: Q. Niu, D.J. Thouless and Y.-S. Wû, Phys. Rev. **B31** (1985) 3372
Maharana and Panda 1991: J. Maharana and S. Panda, Mod. Phys. Lett. **A6** (1991) 163
McCabe and Ouvry 1991: J. McCabe and S. Ouvry, Phys. Lett. **260B** (1991) 113
Moore and Read 1991: G. Moore and N. Read, Nucl. Phys. **B360** (1991) 362

Moore and Seiberg 1989: G. Moore and N. Seiberg, Comm. Math. Phys. **123** (1989) 177
Murthy *et al.* 1991: M.V.N. Murthy, J. Law, M. Brack and R.K. Bhaduri, Phys. Rev. Lett. **67** (1991) 1817
Pais and Uhlenbeck 1959: A. Pais and G.E. Uhlenbeck, Phys. Rev. **116** (1959) 250
Prange and Girvin (1990): *The Quantum Hall Effect* edited by R.E. Prange and S.M. Girvin (Springer-Verlag, Berlin 1990)
Reichl 1980: L.E. Reichl, *A Modern Course in Statistical Physics*, University of Texas Press, Austin (1980)
Schonfeld 1981: J.F. Schonfeld, Nucl. Phys. **B185** (1981) 157
Schiff 1968: L.I. Schiff, *Quantum Mechanics* (McGraw-Hill, New York 1968)
Scott 1970: G.P. Scott, Proc. Camb. Phil. Soc. **68** (1970) 605
Sen 1991a: D. Sen, Nucl. Phys. **B360** (1991) 397
Sen 1991b: D. Sen, Indian Instiute of Science at Bangalore preprint, IISc-CTS-91-9 (October 1991)
Sen 1991c: D. Sen, Indian Instiute of Science at Bangalore preprint, IISc-CTS-91-11 (December 1991)
Shapere and Wilczek 1989: *Geometric Phases in Physics* edited by A. Shapere and F. Wilczek (World Scientific, Singapore 1989)
Shapiro 1970: J.A. Shapiro, Phys. Lett. **33B** (1970) 361
Siegel 1979: W. Siegel, Nucl. Phys. **B156** (1979) 135
Skyrme 1961: T.H.R. Skyrme, Proc. R. Soc. London **A260** (1961) 127
Spielman *et al.* 1990: S. Spielman, K. Fesler, C.B. Eom, T.H. Geballe, M.M. Fejer and A. Kapitulnik, Phys. Rev. Lett. **65** (1990) 123
Sporre *et al.* 1991a: M. Sporre, J.J.M. Verbaarschot and I. Zahed, Phys. Rev. Lett. **67** (1991) 1813
Sporre *et al.* 1991b: M. Sporre, J.J.M. Verbaarschot and I. Zahed, Stony Brook preprint, SUNY-NTG-91-40 (October 1991)
Sporre *et al.* 1991c: M. Sporre, J.J.M. Verbaarschot and I. Zahed, Stony Brook preprint, SUNY-NTG-91-47 (December 1991)
Stern 1991: J. Stern, Phys. Lett. **265B** (1991) 119
Stone 1991a: M. Stone, Int. Jour. Mod. Phys. **B5** (1991) 509
Stone 1991b: M. Stone, Ann. Phys. **207** (1991) 38
Thouless *et al.* 1982: D.J. Thouless, M. Kohmoto, M.P. Nightingale and M. den Nijs, Phys. Rev. Lett. **49** (1982) 405
Ting and Lai 1992: C. Ting and C.H. Lai, National University of Singapore preprint, NSU/HEP/920501 (May 92)
Tsui *et al.* 1982: D.C. Tsui, H.L. Stromer and A.C. Gossard, Phys. Rev. Lett. **48** (1982) 1559
Ursell 1927: H.D. Ursell, Proc. Camb. Phil. Soc. **23** (1927) 685
Vercin 1991: A. Vercin, Phys. Lett. **260 B** (1991) 120
Virasoro 1969: M.A. Virasoro, Phys. Rev. **177** (1969) 2309
Volovik and Yakovenko 1989: G.E. Volovik and V.M. Yakovenko, Landau Institute preprint (1989).
von Klitzing *et al.* 1980: K. von Klitzing, G. Dorda and M. Pepper, Phys. Rev. Lett. **45** (1980) 494
Wen and Zee 1989a: X.-G. Wen and A. Zee, J. Phys. France **50** (1989) 1623
Wen and Zee 1989b: X.-G. Wen and A. Zee, Santa Barbara preprint, NSF-ITP-89-155 (November 1989)
Wen *et al.* 1990a: X.-G. Wen, E. Dagotto and E. Fradkin, Phys. Rev. **B42** (1990) 6110

Wen 1990b: X.-G. Wen, Phys. Rev. Lett. **64** (1990) 2206
Wilczek 1982a: F. Wilczek, Phys. Rev. Lett. **48** (1982) 114
Wilczek 1982b: F. Wilczek, Phys. Rev. Lett. **49** (1982) 957
Wilczek 1990: F. Wilczek, in *Fractional Statistics and Anyon Superconductivity* edited by F. Wilczek (World Scientific, Singapore 1990)
Witten 1979: E. Witten, Phys. Lett. **86B** (1979) 283
Witten 1989: E. Witten, Comm. Math. Phys. **121** (1989) 351
Wu 1984a: Y.-S. Wu, Phys. Rev. Lett. **52** (1984) 2103
Wu 1984b: Y.-S. Wu, Phys. Rev. Lett. **53** (1984) 111
Wu and Yang 1975: T.T. Wu and C.N. Yang, Phys. Rev. **D35** (1975) 3849
Yang and Ge 1989: *Braid Group, Knot Theory and Statistical Mechanics* edited by C.N. Yang and M.L. Ge (World Scientific, Singapore 1989)

Lecture Notes in Physics

For information about Vols. 1–374
please contact your bookseller or Springer-Verlag

New Series m: Monographs

Vol. m 1: H. Hora, Plasmas at High Temperature and Density. VIII, 442 pages. 1991.

Vol. m 2: P. Busch, P. J. Lahti, P. Mittelstaedt, The Quantum Theory of Measurement. XIII, 165 pages. 1991.

Vol. m 3: A. Heck, J. M. Perdang (Eds.), Applying Fractals in Astronomy. IX, 210 pages. 1991.

Vol. m 4: R. K. Zeytounian, Mécanique des fluides fondamentale. XV, 615 pages, 1991.

Vol. m 5: R. K. Zeytounian, Meteorological Fluid Dynamics. XI, 346 pages. 1991.

Vol. m 6: N. M. J. Woodhouse, Special Relativity. VIII, 86 pages. 1992.

Vol. m 7: G. Morandi, The Role of Topology in Classical and Quantum Physics. XIII, 239 pages. 1992.

Vol. m 8: D. Funaro, Polynomial Approximation of Differential Equations. X, 305 pages. 1992.

Vol. m 9: M. Namiki, Stochastic Quantization. X, 217 pages. 1992.

Vol. m 10: J. Hoppe, Lectures on Integrable Systems. VII, 111 pages. 1992.

Vol. m 11: A. D. Yaghjian, Relativistic Dynamics of a Charged Sphere. XII, 115 pages. 1992.

Vol. m 12: G. Esposito, Quantum Gravity, Quantum Cosmology and Lorentzian Geometries. XVI, 326 pages. 1992.

Vol. m 13: M. Klein, A. Knauf, Classical Planar Scattering by Coulombic Potentials. V, 142 pages. 1992.

Vol. m 14: A. Lerda, Anyons. XI, 138 pages. 1992.

GPSR Compliance

The European Union's (EU) General Product Safety Regulation (GPSR) is a set of rules that requires consumer products to be safe and our obligations to ensure this.

If you have any concerns about our products, you can contact us on ProductSafety@springernature.com

In case Publisher is established outside the EU, the EU authorized representative is:

Springer Nature Customer Service Center GmbH
Europaplatz 3
69115 Heidelberg, Germany

Batch number: 10620881

Printed by Printforce, the Netherlands